THE CREATION AND THE EXTINCTION

GLOBAL WARMING AND ICE

Vern G. Rickey

authorHOUSE®

AuthorHouse™
1663 Liberty Drive
Bloomington, IN 47403
www.authorhouse.com
Phone: 1-800-839-8640

First published by AuthorHouse 7/29/2009

ISBN: 978-1-4490-0789-8 (e)
ISBN: 978-1-4490-0788-1 (sc)

Printed in the United States of America
Bloomington, Indiana

This book is printed on acid-free paper.

CONTENTS

CHAPTER - 1
THE CREATION

The beginning started when a great cloud of interstellar gas and debris from the explosion of a super giant star came together and formed an accretion ring or disk of gas, dust, and debris. This event took place in one of the limbs of our spiral Galaxy, the milky way Galaxy. The explosion of the super giant star was a super nova event. The super nova event is so powerful and the energy released from the event is so great , that when the event takes place the exploding star will light up an entire Galaxy. These events are rare in our Galaxy and because of this, members of the Astronomical community view and study supernova in other near by galaxies.

As the accretion disk centralized, most of the debris collected at the point of rotation and formed a beautiful yellow star, that evolved into our sun. The accretion disk degenerated into clouds of swirling gas and rings of super nova debris that formed a series of great flat circular disks or rings much like the rings of Saturn, Uranus and Neptune do today only on the broad scale of the entire solar system. These massive rings of super nova debris and interstellar gas reached completely around the newly forming star. Each of the rings or disks of flowing moving material then sub centralized and formed the planetary system that we are familiar with today. It is believed that there were eight planets that formed from the accretion disk and then tiny little Pluto. Pluto has been down graded from the position as a full blown planet to the lowly status as a planetoid by some. Never the less, eight planets formed from the accretion disk and there are nine if again you count little Pluto.

Planets are different than comets in that they form and follow elliptical almost circular orbits around the sun. None of the existing planetary orbits cross each other except little Pluto which is at the outer most limits of the solar system. A comet has an orbit that reaches far beyond the confines of our solar system. The long elliptical orbit of a comet will penetrate the inner solar system, brush very close to the sun and then travel far out back into deep space. There it will eventually be turned around by the suns gravitational field and then come back in to brush by the sun again. Most comets are a mass of dirty frozen light gasses. The surface of many comets will vaporize as they approach the heat of the sun. When they do, they leave a long wispy streaming tail that always faces away from the sun directed there by the power of the solar wind. Some comets cross the orbital paths of the Planets when they cross the solar system in there approach to the sun, and then again when they leave the sun. Comets can and do strike the planets, and when they do there can be devastating results

In the pioneering days of Astronomy, Astronomers calculated that there should be another Planet where the asteroid belt is today. When a detailed search was made of that part of the sky, none could be found. In their search for this planet they inadvertently discovered some of the asteroids in the region and this in turn led to the discovery of the Asteroid belt.

As the accretion disk of interstellar gas and supernova debris centralized, our Sun and the Planets formed from the material. There are nine known planets in our solar system today if you also count little Pluto, but in the beginning there were ten. A fifth planet from the sun started to form in that part of the sky where the asteroid belt is. That would make the numerical order of the planets, Mercury 1, Venus 2, Earth 3. Mars 4, the fifth planet from the sun 5, Jupiter 6, Saturn 7, Uranus 8, Neptune 9, and little Pluto 10. As this planet has no name and for want of something better to call this planet, I am going to call it the fifth planet from the sun or the rogue planet.

As the sun and planets gained mass by accumulating interstellar gas and supernova debris from the accretion disk, giant Jupiter with its great and ever increasing mass began lengthening the orbit of the fifth planet from the sun. The orbit of the fifth planet as it was forming where the asteroid belt is today, became progressively a longer sharper ellipse influenced by the ever increasing and powerful tidal forces exerted by the Jovian Giant. As the mass of the sun and all the planets in the newly forming solar system increased, great Jupiter increased its mass and tidal force until it placed the fifth planet from the sun into a long elliptical orbital length that crossed the orbit of Mars. As the mass of all the astronomical bodies in the solar system increased, the orbit of the fifth planet finally and fatefully deepened until it crossed the orbit of the earth.

If we draw a map of a near circular orbit like the orbit of the earth is, and then draw a larger narrower ellipse that is out side of this, it can be seen that where the two orbits cross, the orbital paths of both planets are almost traveling in the same direction to each other. This is at two sections of both orbits. This means that both objects are traveling in the same direction and both are on the same orbital line of travel around the sun at these two areas of both orbits. This then is the setting for worlds to collide. After the fifth planet started crossing the orbits of both Mar and the Earth it was just a matter of time until the inevitable happened. After many millions of crossing orbits around the sun, the fifth planet would eventually strike one of the inter planets and it did. It hit the earth.

Needless to say, the fifth planet from the sun did not strike the earth, instead the earth struck the fifth planet. The earth with its inside orbit would be traveling much faster than the intruder. The intruder had a long elliptical orbit that reached clear out toward great Jupiter and maybe even beyond. There is a possibility that the rouge planet may have had a long enough orbit that it reached clear out beyond Jupiter and was turned back by the powerful gravitational field of the great planet.

NASA, when they sent a deep probe to the outer most limits of the solar system used the grand alignment of the outer Planets and the whip lash power of each gravitational field to propel the probe on its journey. This allowed the probe to cross the great distances from planet to planet and record information and images from each in secession. It isn't inconceivable that a similar whip lash like series of events could have taken place if the rogue planet were to have came to close to great Jupiter with its ever increasing mass as the solar system was forming.

As the two worlds came violently together, the fifth planet was slightly ahead of and just a little inside of the earths orbit. It was also just a little higher than the earths equatorial zone. When the destructive force of the earth impacted the intruder, the intruder disintegrated, melted, and then pasted or fused itself to the surface of the earth. When the collision took place, it was slightly inside earths orbit and threw our little world in to a fast spin from west to east. The force of the impact was so great that it repositioned the axis of rotation of the earth over about twenty to twenty two degrees from the planetary plane to where it is today.

The mass of the earth is large enough that the interior of the earth is hot. It is so hot that much of it is in a fluid or Simi-fluid plastic type state. The impact with the intruding planet had a hydraulic effect on the fluid or Simi-fluid interior of our little world, and had sufficient force to blow or eject the outer most ridged tectonic plate material right off the face of the earth into the immediate surrounding space. The collision was not straight on but to the left or inside the orbital line of the earth in its path around the sun. This is what induced rotation to the earth as west to east and at the same time the outer crust plate material that was ejected was thrown into a circular path around the earth that also was moving in a west to easterly direction. Over a period of time a series of broad rings and again much like the rings of Saturn are that we see today developed from this material. These rings of material that were prompted by the persistent tidal forces of the sun then centralized and formed our moon.

Rocks and geological samples returned from the moon have been dated at four point six billion years old. Meteorites that have fallen to the earth have been dated at four point five billion years. Rocks found on the continents of the earth have been dated at three point eight billion years. There are also rocks found on the continents of the earth that have grains of sand like material imbedded in them that have been dated at four point one billion years. This is the oldest material found on the earth so far. The fifth planet from the sun when the earth impacted it was vaporized, disintegrated, and then melted. This

Simi-fluid material then fused to the molten mantel of the rapidly rotating earth and formed one huge Mega Continent.

If the pre-impact outer crust plate material that was blown off the earth and formed the moon had been melted by the impact, then the age of the moon would have been the same as the age of the continents. This is at three point eight to four point one billion years. If the material that formed the moon had been melted by the impact, there is a possibility that the material returned from the moon is instead meteorite material that fell there since its inception.

The moon has no atmosphere to burn up meteorite material when it impacts the lunar surface like the earth does. There is a possibility that the moon has a shell or layer of meteorite material that covers the entire lunar surface and has accumulated since the moon formed. There could be an untold amount of meteorite material that fell there in the last three point to eight to four point one billion years since the creation of the moon. This material would be at four point five to four point six billion years which is the same age of Meteorites because they are Meteorites. This means that the lunar material returned from the moon may have been meteorite material instead of true lunar material. Because of this, the only way to determine the true age of the moon would be to make penetrations into the lunar surface and then date this material. There is also a possibility that material retrieved from a deep impact crater on the moon could be some of the original pre-impact outer crust plate material from the earth that formed the moon. Of course if, the pre-impact outer crust plate material had not been melted by the impact then the age that has been determined to be at four point six billion years would be the correct age of material that formed the moon. The ocean floors of the world are no older than two hundred million years and have been dated as such.

The encounter had to of happened slightly inside the earths orbital line and just a little above the equatorial plane of the earth. This is because the earth is in a fast rational sequence and this sequence is from west to east. If the impact had of been from the side, it would have changed the earths orbital pattern from the near circular orbit that it is in today. The earths orbit would possible be a steeper ellipse. If the impact when the earth hit the intruder had of been slightly to the right and out side its orbital line around the sun then the rotation of the earth would be backwards to what it is today. The rotational sequence of the earth would be from east to west instead of west to east. Also the earth is tipped over on its axis of rotation about twenty two degrees from the orbital plane and the great mass of the continents is predominately in the northern hemisphere. (Europe, Asia, and North America). This places the point of impact slightly above the equator and inside the orbital line of the earths path around the sun.

The most accepted theory of the moons formation other than what has been stated here, simply states that a planetoid or celestial object impacted the earth and not the earth striking the intruding object. The theory also just simply states that the impact splashed material from the earth into the surrounding space and that is what formed the moon. No where does the theory state where the object came from and then what happened to the object after the event took place. This object is supposed to have been at least twice the mass of mars. If it was then where is it now and where did it come from?

Geologists in their study of the earth and the earths crusts again have determined that the oceans floors of the world are only about two hundred million years old and are primarily made of three types of material. The sea floors are rock formations made mostly of Basalt, Gabbros and Sediments. The continents are made mostly of a different lighter type of materials mainly Granite and Diorite. It has been proven that the continental plates have moved independently of each other and they are moving right today. Most geologists believe that the continents came from with in the earth by volcanic action. This is because of the crystalline structure of the continental material. When material cools below the earths surface, it does so rather slowly. Material that cools above the earths surface cools much more rapidly. Because of this the crystalline structure of the same materials are different and a Geologist can tell this. When the earth impacted the fifth planet, the force of the impact fused the molten material of the fifth planet to the molten face of the earth. The material that comprised the fifth planet then would have cooled very slowly because most of it cooled well below the earths surface and what was on the surface cooled more rapidly .

The little red planet Mars has a rotational sequence to it similar to the rotation of the earth and Mars is also tipped about twenty five degrees from the planetary plane on its axis of rotation. The geological strata of the northern and southern hemispheres on Mars is completely different. In the northern hemisphere, Mars has what has been termed the "Thaïs Bulge" and five extremely large volcanoes. The largest volcano found there is Olympus Mons and stands sixteen miles high. This huge mountain is twice the height of Mount Everest, (the highest mountain on earth) and it is the largest volcano in the entire solar system. There are also three other large volcanoes, Arsia Mons, Pavonis Mons, and Ascresus Mons that are all lined up in a row. There are also areas of featureless plains and another great volcano named Alba Patera. This volcano is four miles high and is a thousand miles wide. There is also a huge fracture or rift called the Valles Mariners. This great chasm is twenty five hundred miles long and four miles deep. The famous Grand canyon found here in Arizona is only a mile deep by comparison. The Grand canyon of Arizona is listed as one of the great seven wonders of the world. It would be hard to contemplate such a tremendous chasm here on earth that is four miles deep and twenty five hundred miles long.

The southern hemisphere of Mars is completely different from the north in that there are what appears to be huge impact craters on the surface of which one is three point eight miles deep and twelve hundred and fifty miles across There are several other very large craters plus many smaller craters of which most are in the southern hemisphere. The general surface terrain in the southern hemisphere is higher than it is in the northern.

What kind of forces could have formed such geological deformities as these? Olympus Mons is the highest volcano in the solar system. It is believed that these Volcanoes are what has been termed shield Volcanoes in that repeated eruptions are what built the layers of the cones. It isn't so much the size of the tremendous deformities but that these deformities are so completely different in the northern and southern hemispheres.

When the fifth planet from the suns orbit started crossing the orbit of Mars, it no doubt made close pass bys of the red planet. In fact the two worlds almost collided and maybe many times. This planet with its crossing orbit could have hit Mars instead of the earth and it almost did. A series of extremely close pass bys would have raised the great "Thaïs Bulge" and repeatedly triggered all the volcanoes that are in a line found on Mars today in the northern hemisphere. Other wise the great Mountains on Mars are not volcanoes like we have here on earth. They only became active when there was a close encounter or near miss with the fifth planet in its crossing orbit. These volcanoes are instead evidence of the extremely close pass bys that were made by the fifth planet. Repeated close pass bys maybe centuries apart, created such gravitational tidal forces that they almost ripped the red planet apart. The great stress fracture "Valles Mariners" also in the northern hemisphere that is twenty five hundred miles long and four miles deep is testimony to the tremendous tidal forces created.

Also there are other huge rifts and valleys that have been speculated about as being formed by water. Instead these great valleys and rifts were made by gravitational tidal forces induced by the close pass bys of the fifth planet. The appearance of what seems to be water erosion, was created due to the great sand storms that periodically form on Mars and this is what eroded these huge fractures. This means that contrary to current belief about hidden water below the surface of Mars that may have eroded these fractures, this water erosion never took place. There may be some water on Mars, but not enough to have formed the great fractures found there today. Mars is believed to have a molten interior and an iron core partly because of the presence of the volcanoes.

Just the opposite took place in the southern hemisphere. There are no volcanoes in this region but there are many of what appears at first to be impact craters and the terrain is higher in the southern hemisphere. When our moon here on earth creates a tidal force on the earth, there is a tidal force on the opposite side of the earth at the same time. The solar tidal force is lesser and is usually eclipsed by the stronger lunar tidal force. Because of this there usually are two tides on the earth each day. It is not inconceivable that the high ground found in the southern hemisphere on Mars formed in reaction to the extreme tidal force generated by the close pass bys of the fifth planet. A pass by of any where from a hundred miles to maybe a thousand miles would have devastating effects on the planet Mars. The tidal force was so strong that the terrain of mars in the southern hemisphere was actually bulged by the opposite tidal force just the same as ocean water does here on the opposite side of the earth.

The gravitational force is strongest at the apex of a sphere. At this point the gravitational force between the two planets, raised the "Thais bulge" and the volcanoes in the northern hemisphere. Three of them formed in a row following the path of the intruding planet as it passed by. About the same time, the opposing tidal force created the higher terrain found in the southern hemisphere. The concentrated internal gravitational force at the apex of the sphere created what at first seems to be an impact crater that is really a great sink hole in the southern. Many of what seems to be impact craters are instead sink holes formed when the event took place. The very ground in the southern hemisphere of Mars was actually boiling in response to the gravitational forces generated by the tidal actions. It was boiling not because of heat, but because of the weightlessness of the surface material created by the power full tidal actions.

The rogue planet that was calculated by some to be at two to two and a half times the mass of

Mars that impacted the earth in the formation of our moon then would have been much larger than Mars. This would have subjected the little red planet to untold gravitational stress forming the unique and puzzling geological formations found there today. It is not inconceivable that the little red planet during the pass by was almost completely torn apart by the tremendous tidal forces of the larger body.

Astronomers, Geologists, Geophysicists and Astro-physicists apparently have based there belief that the volcanoes found on Mars are shield volcanoes because of the volcanic actions found here on earth. Volcanoes here on earth build up their cones because of repeated eruptions over a period of time. The volcanoes found on Mars are huge just to say the least and at first it would appear as though this is the only way that they could have attained their tremendous size.

There are mainly two types of volcanoes found here on earth. Most volcanoes are tectonic plate line or fault line volcanoes like those found around the Pacific Rim. Volcanoes like the Hawaiian chain are formed by a hole or vent in the tectonic plate that fuels the volcano as the plate moves building the chain of islands. Almost all volcanoes of any type here on earth build a cone or mountain through repeated eruptions of the volcano. Hence the term shield volcanoes. The interior of the earth is in a molten state and this is what fuels the eruption of the volcanoes. Most of the volcanoes on the west coast of the

American continents are fueled by increased friction as one plate rides up over another.

The point that is being made is that, the volcanoes and geological strata found on Mars have not been fueled like volcanoes found here on earth. There is evidence and opinions that Mars is a rubble planet and has never had a molten interior like the earth has. Because of this the volcanoes and geological strata found on Mars may have been formed by a single pass by of the rogue planet and not repeated pass bys as has been previously stated. The internal movement of material with in Mars because of the powerful gravitational forces would create great frictional internal heat inside the planet. This heat may have been sufficient to induce the interior of Mars into a molten state and this may have been what fueled the eruption of these great volcanoes. Again these volcanoes are only found in the northern hemisphere on Mars and not all over the planets surface.

Of the four interior planets, Mercury, and Venus, have very little rotation where as the Earth and Mars have a great deal. The fifth planet from the sun would have spun the earth into its fast rotation when the earth eventually struck the intruding planet. The fifth planet from the sun in the close pass bys of Mars that almost hit the planet and raised the "Thaïs Bulge," would have invoked the rotation of the red planet that we see today. This could have also tipped the rotation of the little planet over the twenty five degrees to where it is now. The rotation of Mars at the present time is very close to the same rotation of the earth at twenty four hours and twenty seven minutes. There is a possibility that the earth endured several close pass bys made by the fifth planet and already had some rotation to its planetary mass when the fateful encounter took place so many billions of year ago.

The fact that the earth and the fifth planet from the sun were both traveling in the same orbital direction softened the collision and the earth was not destroyed by the event. Instead it led to the formation of the huge mega continent, the moon, and the fast rotation of the earth that we have today. This also means the mass of the impacting planet may have been less than what has been calculated at two to two and one half times the mass of Mars. This is because both of the planets were traveling in the same orbital direction and if the earth already had some rotation induce by previous pass bys of the fifth planet, the union of the two could have been almost leisurely. The fifth planet from the sun may have had rotation itself induced by the close pass bys of Mars and the Earth and this would have softened the impact more yet. Collisions can come from any angle. Head on is the worst, followed by side impact, and then impact when both are traveling in the same direction. Both traveling in the same direction is what took place when the fifth planet from the sun was impacted by the earth.

The Asteroids in the Asteroid belt are what is left of the planetary ring or disk of super nova debris that was forming as the fifth planet. The entire solar system was filled with such debris before this debris centralized and formed the planets. As the path of the fifth planet steepened its fateful elliptical orbit, it started missing the debris that was in a more circular stable orbit. Individual asteroid by them selves do not have sufficient mass to be greatly effected by giant Jupiter but when they centralize into a small planetoid then the Jovan giant exerts its great power. There are other asteroids out side the asteroid belt that have been left over when the solar system formed and there are other planetoids that cross the orbit of the earth. There are Planetoids that right today that make close pass bys of the earth and there was a great planetoid that impacted the earth at the Yucatan peninsula in Mexico just sixty five million years ago. It is believed that this event brought an extinction to the dynasty of the Dinosaurs and Mars has two small moons both of which are probably captured asteroids.

Calculating the orbit of an asteroid at first would seem to be rather straight on and this is routinely accomplished with computers. The sun is used as a focal point in the Kepler calculations and perturbations are calculated when a second body is involved but what about collisions. When the surface of most asteroids is examined, they have a worn and pocked marked like appearance. This could be interpreted to mean that many Asteroid have collided with each other and maybe many times. There are Asteroid that have been found that have gone into orbit around each other. If there were an impact or collision between two or more Asteroid on the far side of the sun that could not be detected then it would be impossible to tell where they would go. It would be bad enough if they were perfect spheres like pool balls but the irregular surface of most asteroids introduces the element of random chaos into their position. This could be part of the reason that, Asteroids are found not only in the asteroid belt, but in different parts of the solar system and in crossing orbits with the earth.

It is believed that the early moon was much closer to the earth and there may have been rotation then to the moon. What rotation there was if any, has ceased and now only one side of the moon faces the earth in perpetuity. The surface of the mangled earth after the catastrophic impact was an exposed sea of raging hot molten lava and the earths entire semi fluid mantel was laid bare like an open wound by the event. What was left of the disintegrated fifth planet after the impact was pasted to one side of the globe with the earth spinning wildly on its axis of rotation.

When this event took place and formed our moon, raises the question of how long was the earth in existence before the inter-Planetary impact between the fifth planet and the earth? The dating of three point eight billion years for the oldest rocks found on the continental crust material could be from the time of the Planetary impact and consequently its melting. The grains of sand like material found in some of these rocks dated at four point one billion years could also be from the time of the impact or maybe something else. They could be from the time of the collision. True! Or they could also be some of the residue from the early pre-impact outer crust plate material that was ejected into the immediate space around the young earth right after the impact. This pre impact outer crust plate material would be older than the continental plate material found on the earth today. This material was part of the original earths crust that was ejected into the surrounding space by the impact. This material formed the moon and geological samples returned from the moon are at four point six billion years. These grains of material could have settled back down over the exposed raw mantel at the earths surface and become imbedded in this hot molten material right after the interplanetary impact took place.

The earth itself and everything in our solar system would have to be four point six billion years old. This would be from the time of the super nova event and consequently the last melting of the material. The dating of the age of the continents could be anywhere from three point eight billion years to four point one billions years and would be from the time of the interplanetary collision with the fifth planet from the sun and the last melting of this material. The time of the age of the ocean floors at two hundred million years is something else that will be take up shortly.

The planet Venus has been referred to as earths sister planet. Venus is almost as big as the earth but not quite. The diameter of Venus is believed to be about seven thousand five hundred twenty miles, where as the diameter of the earth is believed to be about seven thousand nine hundred twenty six miles. Venus as a very bright star can be seen in the evening and in the morning sky and has been the creation of Roman and Greek Gods as a God of great beauty and love. Venus has a mysterious mystic about it as both the Evening and the Morning star. There has been lots of boats and many ships named either the Morning Star, or the Evening Star. If you look at Venus through a small telescope you can see no distinguishing features on the planet at all but you can see the phases similar to the phases of the moon as quarter, half, and full. There the similarity to our earth ends.

The surface of Venus is about as close to being the infernal fiery depths of hell as you could possibly imagine. The surface temperatures there is about eight to nine hundred degrees anywhere on the planet. It is so hot that any lead or tin on the surface would be melted in molten pools. Magnify the shimmering heat that you see on a hot day over an asphalt road about ten times and that is what it would look like on Venus. You and I on the surface of Venus would not last very long. The water vapor of our bodies would instantly explode as a couple of super hot burst of steam. Because of the extreme temperatures there is very little if any water on Venus.

The reason you can see no surface features on Venus is because the planet is shrouded in thick clouds of Carbon Dioxide with Sulferic acid as cloud particles. About ninety six to ninety seven percent of the atmosphere on Venus is Carbon Dioxide. What rotation there is to Venus is in a slow retrograde and Venus is the only planet in the solar system that has a slow rotation that turns backward to its orbital motion. Venus's orbit is inside that of the earths and is much closer to the sun than is the earth that is obvious and for this reason alone would account for much of the heat on the planet. You could call the surface conditions on Venus a super hot example of the green house effect. The thick clouds of sulfuric acid and carbon dioxide atmosphere hold the heat in making the surface temperatures almost the same any where on the planet day or night. Venus is almost as large as the earth an will have and interior source of nuclear heat from with in the planet that is similar to the one we have on earth. The interior source of heat on Venus would have almost no chance of escaping during the night time cycle of the planet. This is because of almost no planetary rotation and the thick clouds of sulfuric acid particles and dense carbon dioxide gasses that make up the atmosphere. This would be a major contributing factor to the extremely hot temperatures found at the planets surface.

About one half of one percent of the atmosphere right now here on earth is made up of the carbon gasses. Even a small build up of the one half of one percent of the atmosphere that are the carbon gasses in our atmosphere, will have a similar effect to our earth as do the clouds of sulfuric acid and carbon di- oxide atmosphere on Venus do. They will keep both sources of heat, the interior source from deep with in the earth, and the exterior source from the suns rays from rising from the surface of the earth and escaping into the outer space around our planet during the night time cycle of the earths rotation. If this carbon gas barrier to the heat escaping becomes to great, then the heat will build up to dangerous levels as can be seen on the surface of Venus. The surface of Venus is a hellishly hot infernal region that mankind can never ever set foot on.

If this analyses should be correct, then it again raises the question of what was our world like before the planetary collision of the earth and the fifth planet from the sun that formed our moon? Was the earth similar to Venus with very little planetary rotation if any induced by a single tidal force of the sun only? Was one side of the earth primarily facing the sun much like Venus and the planet Mercury do today? Both of these inner planets have very little rotation of their planetary mass compared to what the earth and Mars have. The energy of the fifth planet from the sun hitting the earth where it did is what put the earth into its fast rotation that we have now. To support this hypothese, it has been determined that the rotation of the earth is slowing down and the moon is ever so slowely moving away from the earth. This means that some catastrophic event put the earth into its accelerated rotation that we have today and it has been slowing down ever since this happened. Were there maybe pre impact oceans on the earth that were vaporized by the event and if there were, what were they composed of ? Was there an atmosphere that was blown of into the sourrounding space by the impact? There is as much Carbon Dioxide here on the earth as there is on Venus only it is stored in rock formation. Did the early earth before the collision with the fifth planet have very little planetary rotation and a Carbon Dioxide atmosphere similar to the one found on Venus right now? If it did then were there the high temperatures, no water, and surface conditions on the earth then similar to those found on Venus today? These are all valid questions and they may be questions that can never be answered and something that no one will ever know.

It is believed that the moon when it formed was much closer to the earth at this period of time than it is now. This and with the fast erratic rotation of the earth and powerful tidal motions induced by a closer moon were a power full force in the ensuing evolution of our world. The earth rotates from west to east and the moon rotates in its orbit around the earth from west to east. We see the rotation of the earth as an illusion with the moon, planets, sun, stars and galaxies all rising in the east and setting in the west. Astronomers call this motion, right ascension. As the earth spins on its axis of rotation, it passes the gravitational and tidal forces of the moon and the sun approximately thirty times a month or three hundred and sixty five times a year.

This lighter material that would form as a single super continent was literally floating on a sea of molten magma. After the moon formed, the relentless repetitious tidal forces exerted by the closer moon, the sun, and the faster revolving earth, moved the lighter floating material into a slow ponderous motion from east to west. This continental material moving with the accelerated rotation of the earth would catch and pass the tidal forces of the moon and sun twice a day.

Because of this there are four motion induced forces at work on the continents. The first is the lifting tidal action of the moon and the sun on the continents and on the oceans of the world. This is well know and on the coast you can buy a tide book almost anywhere that predicts the tides for the coming year. Approximatel twice a day the tidal forces raise a dome of water that is about sixteen to eighteen inches high that follows the moon as the earth revoleves on its axis of rotation and this is highly predictable.

The force of gravity has two actions to a motion. When the motion is away from the force of gravity as an acceleration away from the earth, the force of gravity increases. These are measured as Gs of force by astronauts at lift off. When the motion is toward the force of gravity as in sky diving, the force of gravity decreases and the sky diver experience a feeling of weightlessness. This motion induced force on the continents is the rotation of the earth itself, revolving on its axis of rotation carrying the continents away to the east from the tidal action excerted by the moon and the sun. This motion increases the tidal action as a gravitational force exerted by the moon and the sun on the continents because of the curvature of the earth. The fact that the surface of the earth is curved changes the tidal force of the moon into a force of acceleration.

This same tidal action on the continents is again a factor as the earthly revolution is completed and the motion of the continents is toward the two tidal actions. Here the force of the tidal actions are reduced by the motion of the continents being carried toward the tidal action excerted by the moon and the sun. Again it is the curvature of the earths surface in motion that promotes the change in the moons tidal force into a force of acceleration.. The moon circles the earth approximately once a month in its orbit and the continents circle the earth approximately thirth times a month with the rotation of the earth. This same tidal actions of the moon and the sun are also believed to be a factor in the slowing of the earths rotation. Venus and Mercury both have very little planetary rotation and both face the sun with very little deviation from this gravitational orientation. Mercury has a pronounced orbital deviation that was predicted by Albert Einstein's General theory of Relativity.

The fourth force is also well known as centrifugal force. Our earth is rotating very rapidly compared to the partial rotation experienced by the other two inner planets. The earth is not a true sphere but is an oblate spheroid. An oblate spheroid is a rotating solid geometric figure of revolution that is flattened at both axis of rotation and broadened at the central great circle of revolution. An oblate spheroid forms when rotation is invoked to a sphere. This means that the earth is flattened at both the north and south poles and is fatter at the equator. This translates into a lifting force of the continents at the equator.

At the time just after the impact of the fifth planet, there was just one great continent on the earth. After the moon formed it started to move and it moved from east to west in response to the repeated lunar and solar daily tidal forces. These tidal forces as they moved over the continent, would not have moved the whole mass of the continent all at one time. Instead, the mass of the single great continental crust stretched and cracked in some places in response to the pressure. The earths crust can and will stretch, crack, and build up tremendous pressures with the ponderous motion of the plates. When it does move, great chasms will open up here and there and then they close. There are instances of earth quakes that have baffled scientists. This is because there were no known fault lines at the point of the quake. In many earth quakes, the movement of the plates that form the earths crust have been measured. In some cases they have moved as much as twenty feet and this can happen in just seconds relieving the pressure. In this way, the great single mega continent stretched, cracked, and inched along like a giant flat wooly worm crawling along over a molten sea of magma, with the western part of the mass ahead one day and the eastern part inching along catching up the next. How many times could this great single continent have traveled clear around the world like a huge curved raft inching along on the subterranean surface of hot molten magma during the three point eight billion years of its existence? There would have been no land marks on the surface of the earth to measure this continental motion. This is because the single great continent was the only point on the globe to measure to or with or from.

There is no way that this can be authenticated as all geological formations made would have been obliterated by the mega continents motion as it circled the globe. The only geological evidence would be the great mountain ranges that reach from Alaska in the northern hemisphere to cape horn in the southern. Also, there are no great mountain ranges along the western shores of Europe and the west coast of Africa.

The Ocean floors are no older than two hundred million years. The continents may be only about three point eight billion years and our moon is four point five to four point six billion years. The asteroids at four point five to four point six billion years are about the same age as our moon. This then is probably the true age of the solar system.

The continents are made of one material and the ocean floors are made of another. The continents are anywhere from three point eight to four point one billion years old. The ocean floors are made of layers of sediment, Basalt, and Gabbros and are only about two hundred million years old. That is a tremendous difference in tectonic plate age. This means that the two American plates and the whole mega continent for that matter, could have been riding up over the pacific ocean floors along the western Americas for maybe three point eight billion years. This has been in response to the gentle yet persistent tidal actions of our moon, sun, and the faster rotation of our world carrying the great mega continent away to the east in its earthly rotational sequence. This constant tidal pressure would have forced the denser pacific oceanic plate back down under the two American continental plates. The denser pacific oceanic plate then re-melts and becomes magma again.

The old idea that the earth and solar system centralized from hot material is losing favor and there is much evidence to support this conclusion. Even so after the impact there would have been no continental or oceanic plates for many millions of years until the earths surface cooled sufficiently for them to reform. What was left of the fifth planet from the sun that now formed the single great mega continent probably would have cooled first. This means that the oceanic plates would have been much thinner offering less resistance to the westward motion of the great mega continent and the re-melting of these plates would have taken place at a much greater rate. This would translate into increased continental plate motion during this period. For an unknown period of time, both of the American plates as well as the great mega continent have been riding up over and forcing down the pacific ocean floors re-melting them. This same action has also been pushing the pacific plate to the west and this in turn is what formed the Hawaiian chain of islands.

As the earths surface cooled, the atmosphere and oceans were created. It may never be know how many times this great single mega continent circum-navigated the globe guided by the persistent and yet repetitious tidal forces of the moon and the sun. The earth remained as a world with one great continent and one great ocean for many billions of years. If my sources of information are correct, about one hundred and fifty to two hundred million years ago, a great split developed from north to south across what was once a single great continent. The European, Asian, and African, continents slowed their westward motion and the two American continents moved away from them forming the Atlantic Ocean. There is information and opinions that the Atlantic ocean may have opened and reclosed more than once and this may very well have happened. If for some reason the American plates slowed there westward motion and the European, Asian, African plates started moving again, it could have closed the Atlantic ocean and reformed the great mega continent. It isn't inconceivable that this event could have happened more than once.

The earth may have had an early atmosphere composed mainly of carbon dioxide before the inter-planetary collision that impacted the earth and our moon was formed. Again there is as much carbon dioxide stored on the earth as there is in the atmosphere of the planet Venus only it is stored in rock formations and not in the atmosphere. Before the interplanetary collision occurred, the early earth very likely had one side of our planet predominately facing the sun with very little if any rotation of the earths planetary mass. Repeated gravitational interactions because of the close pass bys of the fifth planet and then the interplanetary collision is what induced the fast rotation of the earth. A close encounter by Mars with the same rogue planet gave Mars its rotation that it has today.

Carbon Dioxide is an essential ingredient in the life cycle of all living things on our planet. Carbon Dioxide and Oxygen interact. With out Carbon Dioxide in the metabolism of plant life, all animal life on the earth would become extinct. The survival of all animal life in existence on the planet earth depends solely on the survival of plant life.

No one is going to pretend to explain how life formed least of all me but form it did. Both plant life and animal life came into being and both evolved until the present day. From a religious point of view, God created the heaven and the earth and maybe he did, I do not know. I firmly believe that there has to be some great entity that brought everything into existence. This is simply because we are here but what or who a creator could be is still a mystery to me. What can not be argued with is simply that both of these life forms as plant and animal life are in existence. Another thing is that there is no religion that includes plant life as part of their religion. Still it can not be argued with that plants are living things. They grow thorns and spines as do most cactus and many plants and trees. They do this to protect them selves from people and the animals. They also develop poisonous toxins as poison Ivy and poison Oak do, to again protect them selves from people and animals. There are many poisonous plants, trees and different types of poisonous vegetation all over the world. Neither can it be argued with that plant life metabolizes carbon dioxide and excretes oxygen as a waste product. Also Animal life metabolizes oxygen and excretes carbon dioxide as a waste product. One thing is self evident, and that is that one depends on the other, and the other depends on the one. Mankind with our great brain and superior intelligence has got to come to terms with this fundamental and yet very simple and important truth.

Because of this it would seem as though, if the fifth planet from the sun or it could be called the rogue planet had not been impacted by the earth and our moon formed, the earth that we are familiar with today, would be more like Venus is. It would have a carbon dioxide atmosphere and hardly any rotation to its planetary mass. Surface temperatures would be in the six to eight hundred degree range. Other wise, out beautiful little green and blue planet earth would be more like the hell hole that Venus is to day and uninhabitable by either of the two life forms that is the foundations of our existence. If this would have been the case, then life as we know it could not have formed here on the earth. We would not be here. This means that every living thing on the face of the planet earth since its inception both PLANT an ANIMAL, would never ever even have existed. That in itself is hard to come to grips with and harder yet to understand.

Hurricanes, Monsoons, and Cyclones are great seasonal and revolving storms that act as mixers of the atmospheric gasses of our planet. In order for these great storms to form, shallow open areas of the ocean at certain places adjacent to the equatorial regions of the world are heated by rays from the sun. The planetary rotation of the earth in both the northern and southern hemispheres induces rotation to the air mass over the warm waters of the ocean and the air mass rises. As the warm air rises due to the heat from the warm waters, a large air mass over the oceans surface in the immediate area starts to rotate with the rising air. This air mass then spirals inward toward the rising mass of warm air. When this happens, a tropical depression or tropical storm is created. For Hurricanes, and Cyclones to form from tropical depressions and squalls, requires vast stretches of open warm ocean to gain strength as they move across the water. These storms as they gain strength over the warm water are drawing heat from the water. This in turn cools the water. If the revolving storms did not move across the water to fresh warmer water, it would not gain strength.

Hurricanes, Monsoons and Cyclones only form in certain regions of the world. There are no Hurricanes that form in the south Atlantic ocean and the great Monsoonal weather patterns only form in the Indian ocean. There are Monsoonal weather patterns that form in other regions of the world but they are nothing like the great Monsoons off the coast of India in the Indian ocean. If there were no great fetches of open warm water (80 degrees or more) to feed the Monsoons and tropical depressions after they form, then these geat revolving storms would remain as tropical depressions and squalls. They would not gain sufficient strength to be of any great concern.

We start by pushing the north and south American Continents back against Europe, Asia, and Africa. This in turn would flatten the great western mountain ranges of both of the American Continents. This action would also rejoin the American continents with both the European and African continents as one single mega continent. . We continue to push and push this whole mega Continent back against Japan and the African Continent against Madagascar. Bring Australia, Antarctica and all the islands, back up to eastern Asia and then finally pull India back out into the Indian ocean. By doing this, you would flatten the Himalayas. This in effect would place the continents back together where they came from as the great mega continent that first formed after the fifth planet from the suns catastrophic impact with the earth.

It is interesting to note that when we do this, we may have removed most of the great fetches of open warm ocean water that fuels the Hurricanes, Cyclones, and Monsoons of the world that we have today. We have also leveled most of the great mountain ranges including the great mountain ranges of both of the western American continents and the great mountain ranges that form the Alps and Himalayas of Europe and Asia. By moving both of the American continents east back up against Europe and Africa, the Atlantic ocean would in effect be closed and the great mountain ranges that stretch from Alaska in the northern hemisphere to cape horn in the south would be flattened. This would change all the weathers systems from what they are today for the entire world.

Hurricanes can not form over land, only warm water. The great fetch of warm water that is required to fuel hurricanes across the Atlantic ocean would be eliminated. By moving India back out into the Indian ocean, the great fetch of warm water of the Indian ocean that fuels the Monsoons would be eliminated and by flattening the Himalayas the seasonal high and low pressure cells that triggers the Monsoons would be eliminated. And finaly by moving Australia, Antarctica, and all the islands back against eastern Asia, the great fetches of open warm waters of the eastern Pacific would be eliminated. This means that when the earth had only one mega continent and none of the great mountain ranges that we have to day, there is a possibility that there were less of and maybe even none of the great monsoons and revolving storms that we have to day. These storms are great mixers of the atmosphere as they push warm air up into the atmosphere and even up into the stratosphere.

The carbon gasses are heavier than air and will tend to settle down towards the ground if undisturbed. This can be seen in many major cities all around the country. Although the carbon gasses are odorless and colorless, still the particles that accompany them in the air can be seen. The city of Phoenix here in the state of Arizona is a great city that suffers from yellow air when conditions become calm. This yellow haze hanging in the air over the city can be see as you come into the suburbs. Los Angles, Seattle, and many of the eastern cities, in fact almost every major city on earth has had air pollution problems at one time or another. Beijing, another great city in the republic of China was plagued with air pollution problems just before the summer Olympic games in two thousand eight.

CHAPTER - 2
THE ATMOSPHERE AND
GLOBAL WARMING

The main composition of the atmosphere of planet earth is mostly nitrogen. This gas is about seventy eight percent of the total atmospheric gasses and constitutes the main bulk of the atmosphere.. The other major gas which is oxygen, constitutes about twenty one percent of the total atmospheric gasses. The combination of these two gasses, nitrogen and oxygen, make up close to ninety nine percent of the total bulk of the atmosphere. The remaining one percent is a combination in very small mounts of all of the different gasses that make up the total mixture of gasses that is the atmosphere. Not only is oxygen a major gas of our atmosphere but it is a dominate gas in the form of molecules and compounds. H2-0 as water (Hydrogen two and Oxygen one) covers much of the earth. It has been determined that there are approximately three hundred forty million cubic miles of water on our planet, both salt and fresh. Three hundred forty million cubic miles of water!! Not cubic feet or meters, but CUBIC MILES. That is a fantastic amount of water. At any given time there are about two thousand nine hundred cubic miles of water in the atmosphere as water vapor. There are also about two million cubic miles of fresh water as ground water on the continents.

The gases that make up the remaining one percent of the atmosphere are Argon, CARBON DIOXIDE and small amounts of other gasses. These other gasses are trace amounts of Methane, Helium, Neon, Hydrogen, Krypton, Xeon, CARBON MONOXIDE, and Ozone. It is believed that it has taken about four and one half billion years for the atmosphere to have evolved to its present combination of gasses but this amount of time maybe controversial as the continents maybe only about three point eight billion years old. If the impact of the fifth planet from the sun with the earth was at three point eight billion years then what atmosphere there was before the impact could have been changed by the violent event. Volcanic eruptions have been a major contributor to the formation of atmospheric gasses here on earth. This is true but, I want to stresses that it is less than HALF OF ONE PERCENT of all the atmospheric gasses of the Atmosphere that are the carbon gasses. Now Let that sink in. Less than HALF OF ONE PERCENT OF ALL THE ATMOSPHERIC GASSES ARE THE CARBON GASSES. That is just a tiny small fraction of the total atmospheric gasses that blanket the entire face of the Earth.

Global warming has been proven to be caused by an excessive build up of the carbon gasses in the atmosphere. The carbon gasses, are CARBON MONOXIDE and CARBON DIOXIDE and to a lesser extent METHANE. These gasses have the unique ability to reflect heat that rises from the earth back down toward the earth. It acts something like a blanket around the earth or a one way mirror. The light from the sun penetrates the atmosphere until it hits the surface of the earth. When photons of light hit the surface of the earth they then becomes heat. This heat then rises back into the atmosphere. With out the excessive carbon gasses in the atmosphere, this heat would radiate back out into the space around the earth during the night time cycle and the earth would stay cool. GLOBAL WARMING would not be a problem. With the excessive carbon gasses in the atmosphere, the heat builds up and is contained because it is reflected back down toward the lower atmosphere and the surface of the earth by the carbon gasses. With a build up of large amounts of the CARBON GASSES in the atmosphere, some of the heat from the sun is trapped in the lower atmosphere and on the surface of the earth. It builds up and up with no where to go and because of this, the climate of earth is changing radically. That will be bad enough but that is not where the real concern should be. The real concern and danger that will come from global warming should be the process of evaporation that will accompany the raising temperatures.

All you can see right now because of global warming are some changes in the climate. So it gets a little hotter in this place and a little colder over there, so what. The polar caps are melting and there's some flooding along the coast and some people will have to move to higher ground. Glaciers all around the world are melting. Everybody loves the good old summer time when it is warm and comfortable so we could use a little more heat. Big deal. Believe me that's nothing compared to what is going to happen. That's not where the real danger is in global warming. Moisture from evaporation into the atmosphere is the real danger and should be of great concern. Moisture, as water, moisture as vapor, steam, clouds, mist, fog, rain, snow and finally ice.

Global warming is simply that the earths average temperature is raising, and because of this the earths lower atmosphere is heating up. It is heating up because as I just stated the carbon gasses in the atmosphere are reflecting heat that raises from the earth back down into the lower atmosphere. The average temperature of the earth is heating up, but the earths surface temperature varies widely all over the entire globe. Not only this but the layers or you could call them zones of the atmosphere are all at different temperatures.

The Troposphere which is from the surface of the earth at sea level up to about ten miles is where most of the earths weather takes place. Sun light will penetrate the earths atmosphere and when it does, it strikes the surface of the earth. When it strikes the surface of the earth is it is converted into heat. This heat then rises through the atmosphere and is dissipated into space. Temperatures in the Troposphere decrease with altitude because of the rising heat. Because sun light when it strikes the surface of the earth is converted to heat, the valleys are warm and the mountains are covered with ice and snow that varies with the seasons.

The Stratosphere that is above the Troposphere reaches from about ten miles up to about thirty miles. This is where commercial air craft like to fly because they are essentially above the weather found in the lower ten miles of the Troposphere. The Mesosphere which is above the Stratosphere starts in at about thirty miles and extends up to about sixty. Sixty miles and above is essentially space by some definitions even though there are two more atmospheric zones or layers above this. These two more zones or layers of the atmosphere are, the Thermosphere and then the Exosphere which reaches out to about two hundred miles. The Ozone layer that blocks harmful ultra violate radiation from reaching the lower atmosphere is in between the Thermosphere and the Exosphere. The Thermosphere is a hot zone or layer and temperatures in this zone can reach as much as twenty two hundred degrees F.

The north arctic polar cap is melting faster than the southern ant-arctic polar cap. The earth is tilted on its axis and makes the four seasons as you and any high school kid should know. In the northern hemisphere the earth in its annual orbit around the sun is farther from the sun in summer so we have mild summers in the northern hemisphere and closer in winter so we have milder winters. The opposite is true in the southern hemisphere. They have warmer summers and harsher winters. Because Antarctica is farthest from the sun in winter, it is the coldest place on earth.

When the sun shines on the earth, the rays are steepest in the tropics and slanted more at the poles. The effect is something like the sun shinning on a solar panel. If the panel is turned direct to the sun, it gets the full benefit of the light and works the best. If the panel is tilted so the rays hit at an angle, the efficiency of the panel is greatly reduced. Turn it completely flat to the suns rays and the panel goes almost completely dead. The earth is round and heat from the sun deceases with the curvature of the earth. The farther north and south you go from the equator the more there is a pronounced drop in temperatures. For this simple reason alone the tropics of the earth are much warmer than either of the Poles.

The interior of our earth is also hot. According to calculations it should have cooled down eons ago but has not. The scientific community because of this has come to the conclusion that there has to be nuclear reactions that are taking place deep with in the earth that keeps the interior of the earth hot. Ice land is a volcanic region high in the northern hemisphere where surface temperate are very cold yet extreme volcanic heat is just right below the surface. The crust of the earth is mostly rock and mineral only about three or four, to forty seven or eight miles thick on the continents and maybe about nine, ten to fifteen miles thick in the ocean floors any place on the earth. This thin layer of material is in pieces floating on the hot interior of the earth like loose broken pieces of egg shell. These are the continental and oceanic plates that we live on and that were discussed earlier. The study of the movement of these plates is called plate tectonics. Where the edges of the plates come together, is where most of the major earth quakes on earth take place. Members of the scientific community have extrapolated the tectonic plate motion back in time in order to better understand what conditions were prevalent in different periods of the earths historical past.

The interior heat of the earth itself radiates out to the surface in all directions from with in. So as the earth rotates on its axis, not only heat from the sun in the day time part of the cycle, but the interior heat of the earth is lost from the atmosphere every day during the night time cycle. All of the small amounts of the gasses of the atmosphere excluding nitrogen and oxygen, make up only about ONE PERCENT OF THE TOTAL ATMOSPHERIC GASSES. So in reality, just a small increase of the less than one half percent which are the carbon gasses, will change the ratio of carbon gasses to oxygen and will have a dramatic effect on the climate of the earth.

An increase in the overall heat of the earth will increase the evaporation of a lot more water. When water evaporates it has a cooling effect when it evaporates. A swamp cooler is a devise that uses evaporation of water to cool. It is basically a matt in a box with a pump that pours water over the matt and a fan that evaporates the water from the matt and it cools. A swamp cooler is not refrigeration as in air conditioning where the evaporation of a liquid is contained in a closed unit, but is the direct evaporation of water in the atmosphere. A good swamp cooler can cool a home as much as twenty degrees during the summer when the humidity is low. Swamp coolers are used all over the world in tropical regions to cool homes. This is what is so deceiving about global warming. The cooling effect of global evaporation will hide the heat build up until the water is evaporated. After the water is evaporated is when the heat will really build up and become a serious problem

Just to understand what happens a little better, if you place a container of water out in the hot sun, the water will evaporate in a short time. The container and all around the container will stay cool in the sun light as long as the water is evaporating. Once the water is evaporated, the container still in the direct sun light will build up heat and get very hot. In fact the container will get so hot that you can hardly touch it. If the container is of a dark color, it really gets hot. It is the process of evaporation of the water that keeps the container cool until the water is gone.

Ice sheets and glacial deposits on both of the polar regions of the earth, north an south are melting that is obvious. Just this year in 2008, over a trillion tons of fresh water ice have been lost from Greenland alone due to global warming. What is not so obvious is that there are millions of square miles of open water in the oceans of the world. Not thousands, but millions. The earth is covered with water, two thirds of it by the oceans.

Just to get a rough idea of how big this is, do a little simple arithmetic. It is 25,000 thousand miles around the earth at the equator. Make a strip two thousand miles wide at the equator around the earth. 25,000 times 2000 is fifty million. Take two thirds of this that are covered by water and it comes to over thirty million square miles of water and you know that there is a lot more ocean than that. The surface area of all the oceans of the world will run in to MANNY, MANNY, MILLIONS OF SQUARE MILES OF OPEN WATER. Not only the oceans but lakes, rivers, land, people, animals, and vegetation. ALL EVAPORATE WATER AND GET DRYER. As it heats up, the land gets dryer, and fires are more easily started.

This has happened in the state of California and all the western states right now, in 2008. California alone has had about eighteen hundred forest fires burning out of control. The governor of California made a statement on the news that they used to have a fire season only in the fall when it got dry, now they have a fire season all year round. There were also large fires burning out of control in the state of Arizona and many of the western states, and how many more are burning out of control all over the world that you do not hear about. Australia had a terrible time with record temperatures of over a hundred and twenty degrees and great fire storms that killed over two hundred people. True the fires are believed to be arson started but still the record heat and extremely dry conditions with high winds are what made the fires possible in the first place and made them so terrible.

When conditions are extremely dry, lightning when it strikes can and easily does start forest fires. Conditions are extremely dry because of the increased heat due to global warming. Because of the higher heat, water is more readily evaporated from everything including vegetation. Vegetation is not only trees, grass and plants, vegetation is the wood frame work of your house, your furniture and even most of your food . You do not notice the build up of heat increasing, because of the water evaporating that keeps things cool. Dryer conditions on the land promote violent hot dry thunder storms with an abundance of lightning strikes and very little if any rain. Lightning strikes the ground and starts forest fires, but there isn't enough rain to put them out.

As the earth gets hotter and the land gets dryer, where does all the water go? Trees in a forest are mostly made of water. When the tree burns, all the moisture that was the structure of the tree goes into the atmosphere along with a lot of CARBON MONOXIDE. All that is left of the tree after the fire is a small amount of ash on the ground. If you want to know where all the water went? Just look up in the sky and look at the clouds. Clouds hold cubic miles of water. On a Global scale, under current conditions as stated earlier, the atmosphere holds about two thousand nine hundred cubic miles of water.

This is fundamental, if you heat water it turns to steam and vapor as it evaporates.

This happens all the time. It is what makes the worlds weather systems work. Water evaporates from the oceans of the world, is cooled and falls on the land as rain and snow. The only problem here is if you heat water hotter, it turns to steam and vapor faster. If you heat a lot of water hotter you get a lot of water vapor faster and more of it. So as the earth heats up, there will be much more water vapor in the atmosphere than there has been in the immediate past. The evaporating process of the water will keep things cool until it is to late. When most of the water in certain areas of the continents evaporates it forms desert regions and that is where the real increase in global warming will take place.

The oceans of the world will not all evaporate into the atmosphere of course not, but there will be a lot more moisture in the atmosphere than there has been. When it is real hot, you can not see the water vapor in the atmosphere but it is there as high humidity. When it gets colder it forms Clouds, Mist, and Fog Banks. If it gets cold enough, it condenses back into water and falls as rain. If it gets colder yet, it snows.

CHAPTER - 3
WHAT REALLY CAUSES
GLOBAL WARMING

Now what really causes GLOBAL WARMING and how does all the carbon gases get to into the atmosphere anyway? This is something that most people would not at first think about until they understand what it is, and then they see that it is true. But remember, from a countless number of rain drops comes a great and terrible flood. One of the great contributors to Global warming is you and your wife or your husband, your dog, cat, pets etc. There are approximately SIX AND ONE HALF BILLION PEOPLE ON THE EARTH RIGHT NOW AND EACH AND EVERY ONE OF THEM IS BREATHING IN OXYGEN AND EXHALING CARON DIOXIDE. This means that you, your wife, or husband, your pets, and the entire human race are contributing to GLOBAL WARMING JUST BY YOUR BREATHING ALONE. You do this twenty four hours a day, seven days a week, fifty two weeks a year, for your entire life non stop. The amount of CARBON DIOXIDE THE HUMAN RACE puts into the atmosphere none stop every year by just breathing alone is staggering.

If our breathing contributes to global warming then what about all the animals? In just the united states along there are millions of feed lot beef, dairy cattle, pigs, sheep, chickens, turkeys, etc. we raise for food, that breath the same atmosphere in and out that you and I do every day. With the single breath of every one of these creature, they remove just a little oxygen and exhale a little carbon dioxide. These numbers increase many fold to billions of domestic animals, chickens, pigs, cattle, and what not on a world wide scale that all of humanity keeps and raises just for food. The body mass of a 1000 lb beef is about five to six times that of a typical human being. How many cattle are there on earth? A billion, two billion, five billion. I don't think any one really knows. How many full sized domestic animals which are beef, swine, chickens, turkeys, sheep, ducks, geese, etc., does a typical human being consume in the course of a life time. There are so many that you would not believe the number even if you knew

There are also many millions of animals that we keep as pets, dogs and cats etc. This is not only here in the United States, but people all over the world keep all kinds of animals as pets. I have one myself, a cute little dog and I think he's great. These animals do the same thing that you and I do. They breathe in the atmosphere, and when they breathe out the atmosphere they exhale CARBON DIOXIDE. The amount of CARBON DIOXIDE exhaled by all the domestic animals that we raise and consume as food, and the amount of CARBON DIOXIDE exhaled by all the domestic animals we keep as pets world wide, will exceed the amount of carbon dioxide exhaled by the six and one half billion people of the entire human race many times over.

What about all the wild animals, Deer, Elk, Lions, Tigers, Rats, and Mice, Grizzly Bears Big Horn Sheep etc.? HOW MANY wild animals are there on the earth. Every nook and corner of our planet holds wild animals. They live in every country and on every continent on earth. They live on mountains and mountain tops, in the forests and on the plains. They burro into the ground, hide in the bushes and live in trees. Yes they also fly in the air. Birds!! How many birds are there on earth. Billions!!, trillions!! All these wild creatures all over the earth, breath in oxygen and exhale CARBON DIOXIDE just the same as you and I do. The number of wild animals on our earth is uncountable and that is not all.

What about the air breathing creatures of the sea? Whales, dolphins, seals, Sea otters, walruses and what not. They to breath the same atmosphere as you and I, taking oxygen out and excreting carbon dioxide. When you get right down to it , the planet earth is not nearly as big as you may think it is. There are two major life forms, plant and animal life that have to have a balancing effect to each other and the ratio of carbon gases to oxygen in the atmosphere. One absorbs the carbon gases from the atmosphere and excretes oxygen as waste, and the other absorbs the oxygen and excretes carbon dioxide as a waste.

Forest fires were mentioned before and yes they are of a major concern. Many millions of square miles of forests each year are burned in forest fires all over the world. Over half a million acres of forest have been burned in the state of California alone right now, and the fires are still burning out of control as this is being written. How many other places on earth are there forest fires burning, raging out of control. Look at the millions of tons of CARBON MONOXIDE that are put into the atmosphere in just forest fires alone every year. Not only here in the United states but on a Global scale forest fires are putting BILLIONS OF TONS OF CARBON MONOXIDE into the air and all the moisture that is in vegetation as well. As Atmospheric temperatures accelerate and it gets hotter, fires will be more easily started than they are now. The dryer the earth gets with the increased temperatures, the more easily they will start. Where does all the moisture go when the land dries out from the increased temperatures? Right straight up in to the atmosphere as first high humidity and then as water vapor along with all the carbon monoxide that was the structure of the trees.

Industry and the burning of fossil fuels has been blamed for the increase in the average temperature of the earth that has been referred to as "Global" Warming. I believe that global warming would have happened anyway only at a much more greatly reduced rate. It maybe would have taken many thousands maybe even millions of years for the change in the atmosphere that is the cause of global warming to become the problem that it is today with out the burning of fossil fuels.

The real problem is not so much the extra carbon gasses that we are putting into the atmosphere, but that there just is not enough vegetation left to remove these gasses as fast as we are putting them there. Also we put them in places where the plants can't get to them like in the upper atmosphere. We have destroyed so much vegetation by logging, burning and eating it that there just is not enough of it to support the existing animal life which includes you, me, and every living breathing creature on the entire face of the earth. Industry and the burning of fossil fuels has only compounded a very serious problem and made it worse. The burning of fossil fuels has turned the situation into a red letter screaming do or die emergency. We burn fantastic amounts of Oil, Gas, and Coal.

One coal burning power plant burns two train loads of coal every day and the trains carrying the coal are a mile long. This is the amount of coal that is burned at this plant every single day. This single plant releases one hundred million tons of carbon gas every year and there are thousands of coal burning power plants in the United States alone. Not only the United States of America, but countries like China, India, the Soviet Union, and all the developing nations are burning increasingly greater amounts of fossil fuels primarily coal. The Automobile, Trucks, Trains, and ships all burn millions of gallons of fuel each day and place millions of tons of CARBON

GASSES into the atmosphere. How many ocean going ships at sea are there right now out there burning untold millions of gallons of diesel fuel putting carbon monoxide right straight up and directly into the atmosphere?

And what about Aircraft and the Jet engine? This is one of the greatest of all problems and the worst of the worst. Commercial aircraft utilizing the jet engine as a source of power are seeding the very upper atmosphere where it does the most damage with CARBON MONOXIDE. This is one of the major gasses that traps the heat that is destroying our environment and our existence. They are putting billions of tons of carbon gasses every year into the UPPER ATMOSPHERE (Stratosphere) ON A GLOBAL SCALE. All you have to do is look at the sky and the jet trails that are there. You never used to see to much of them some years ago. Now they crisscross the sky over, making it look like a huge checker board. The efficiency of the jet engine is supposed to have increased burning less fuel, yet there are more and more jet trails in the sky coming from airplanes at high altitudes. What you are seeing as the jet trails in the sky, is the added moisture in the atmosphere reacting to the heat and carbon monoxide produced by the jet engines of the aircraft at greater altitudes. The only thing is, when these gasses are deposited at these altitudes by these machines, there are no plants, trees, or vegetation of any kind up there to metabolize them back out of the atmosphere.

I do not believe that Mankind fully realizes or understands the seriousness of what we are doing by the use of jet aircraft. How in the name of creation or anything else for that matter, are we ever going to remove these gasses once we have mixed them into the upper atmosphere? The answer is we can not. The best we can do is stop it and stop it right now because if we do not, the price we are going to pay will be beyond description.

Do you know where this is leading? As we pore TRILLIONS of tons of CARBON GAS into the atmosphere each day by our breathing and combustion, the earth will get hotter and hotter. Evaporation will get worse and worse with higher and higher humidity at higher altitudes as the heat raises. Hundreds to thousands more cubic miles of sea water from the millions of square miles of open tropical ocean will evaporate because of the increased heat. The entire world will turn into a global sized sauna bath. I repeat, there are THREE HUNDRED FORTY MILLION CUBIC MILES OF WATER on the earth. Like slowly heating up the water in a steam boiler it will take a little time to heat up that amount of water but when it does, evaporation will increase dramatically.

The natural cycle of evaporation and precipitation that the earth has experienced for the last ten to twelve thousand years since the last ice age will be grossly accelerated. The increased moisture in the atmosphere will form more vapor, mist, fog banks, and cloud cover where it is colder. Where is it colder?

It is colder over both of the polar regions of our little planet where the direct light from the sun does not fall.

First the lower atmosphere will become warmer as it is doing now although there are still clear skies at both of the polar regions. As the atmosphere heats up sun light can still reach the ground and this will melt glaciers around the world. It will also melt the ice that has been frozen and stored at both of the poles for thousands of years. This is happening right now in Greenland and Antarctica. As this ice melts, it will enter the worlds water systems. Then as the earth heats up more yet, the evaporation of sea water from all the oceans of the world will become more grossly accelerated. This in turn will produce more cloud cover and this will accelerate more precipitation. When the increased moisture cools over both of the polar regions then it can be seen as heavier thick cloud cover, mist, and dense fog banks. THESE CLOUDS, MIST, AND FOG BANKS, WILL BLOCK THE SUN LIGHT FROM REACHING THE SURFACE OF THE EARTH. This is what will shield the surface of the earth from the sun light reaching the ground at the polar regions where it would turn into heat. The surface of the earth there will then become colder as the heat continues to rise from it, although heat in the atmosphere and in the tropics will continue to be accelerated to higher levels by the containment of the carbon gasses. The increased moisture in the tropical regions where it is warmer, will be as high humidity that can not be seen and this will allow the direct sunlight there to fall on the waters of the worlds oceans increasing evaporation..

This increase in moisture in the atmosphere will come as heavier rain, snow and ice at first. Already the mid west has had record breaking floods. First they had what they called a one hundred year flood of the Mississippi river drainage. Now the floods they have had this year in 2008 are being called a five hundred year flood. What will they call the next big flood, a thousand year flood? Because of the increase in the carbon gasses in the atmosphere these floods are going to get an awful lot worse than that and this is going to happen very soon. Every year in many places on the earth there will be record breaking floods.

Right now in the last few days of March the year of two thousand and nine, the entire state of North Dakota is experiencing record breaking floods. Especially hard hit is the city of Fargo North Dakota. The people of this state are fighting in a valiant effort to avert the red river from flooding their city.

Photons of light do some strange things that Physics can not really explain. They sort of understand what happens but not really. It used to be believed that a photon of light knocked a valance electron out of a metal plate, but because of the electrolyses effect this has been modified to mean that a photon generates an electron. Just how a photon generates an electron is not fully understood by anyone. Anyway this process takes place in solar panels, CCD cameras, and the photovoltaic effect.

A similar event takes place on cloud tops and to a lesser degree on the ground. When sunlight can not reach the ground and become heat because of the cloud cover, the photons generate electrons on the cloud tops. When there is an accumulation of electrons on the cloud tops and in the clouds, the electrical energy becomes lightening. Lightening can strike from cloud to ground, from ground to cloud, and from cloud to cloud. When enough electrical energy accumulates in a cloud, that and the moisture in a cloud becomes a low pressure area or a storm. The surface of the earth in the polar regions will get colder because sunlight can not reach the ground level because of the increased cloud cover that is there for prolonged periods of time. The longer the ground is shielded from the sun light what heat that is there in the ground continues to rise and the ground will get colder and colder.

The atmosphere itself in the polar regions and in the tropics of the world will get warmer. This will translate into much larger, bigger, wetter, more powerful storms originating in the polar regions. These storms then are carried by the earths weather patterns into the temperate regions of both hemispheres. There already have been reports to that effect. Washington state and the Seattle area are experiencing record cold and snow, heavy rains and flooding. The city of Spokane in eastern Washington state has had over five feet of snow so far in this winter of

2008 and 2009 and the winter has just begun. This is an all time record for snow fall in this little city in eastern Washington state. Not only Spokane but the entire nation from coast to coast has had record breaking snow and ice storms with record breaking temperatures.

There are those that claim that GLOBAL WARMING is a hoax because of the increase in colder longer storms in winter. The truth of the matter is, it is getting colder at the earths surface in the polar regions because of the increased cloud cover blocking the sun light from reaching the ground and generating heat. The atmosphere itself is getting warmer because of the same increased cloud cover that holds the heat in the atmosphere. When the sun light does reach the ground in the polar regions, the white snow and frost covering the ground from the increased precipitation will reflect the light and the heat back up into the atmosphere. The accelerated evaporation of ocean water in the tropical regions will accelerate precipitation in the north and south polar and temperate regions. It will start snowing heavily as it has in eastern Washington State, beginning at the North an South poles. When the accelerated evaporative process reaches a certain saturation point, IT WILL START RAINING AND SNOWING AND IT WILL NOT STOP. THE RAIN AND SNOW WILL FORM ICE STORMS FREEZING THE WARMER RAIN WATER AS IT MAKES CONTACT WITH THE COLDER SURFACE OF THE EARTH. There has been an increase in the intensity and frequency of ice storms in much of the eastern

United States in just this winter of two thousand eight and two thousand and nine alone.

Look at any map of the world. The greatest land masses on earth are North America, Europe, and Asia and are in the northern hemisphere. The greatest land masses in the southern hemisphere are the lower part of south America, the lower part of Africa and all of Australia and Antarctica. When it snows over land, the snow sticks and builds up. When it snows over the ocean, the snow falls into the sea and is melted. When it snows in the northern hemisphere with the great land masses, (North America, Europe, and Asia) the snow will fall on and then stick to the land. When it snows in the southern hemisphere where there is mostly sea, the snow will fall into the sea and be melted.

The tropical regions of the earth will get hot and get hotter. As the heat increases, evaporation will keep things cool until the moisture content of the land decreases. Like the container of water in the sun, after the water is gone, the heat will increases dramatically forming large waste land deserts. Right now Australia that great country in the southern hemisphere is experiencing increased drought, record heat, and loss of water. Because of the increased heat in the tropical regions, the increased moisture will remain as high humidity even up to higher altitudes. The sun light will still be able to go right straight through the atmosphere and be able to reach the ground and the sea, building up more heat. This is how the heat will reach the oceans of the tropical regions of the world and heat then up more, accelerating evaporation. This along with the internal heat of the earth being contained will push temperatures to higher and higher dangerous levels. The ocean levels on a global scale will not increase as currently believed because of the polar regions melting. They very likely will increase some at first but as temperatures increases they will instead recede because of the extreme evaporation that will take place from a warmer earth. In the last ice age, the ocean levels dropped by as much as four to five hundred feet. If this happens then New York city, San Francisco, Seattle, London, all the great seaport cities all over the world will become high and dry. This alone would cripple trade and commerce all over the world.

There have been several ice ages in the past. There is also a theory of Galactic rotation that tries to explain the ice ages. There is not an Astronomer or Astro—Physicist on the face of the earth that can explain the shape and rotation of a spiral Galaxy let alone the more complex shape of a Barred spiral such as the Barred spiral NGC 1365 in the Fornas cluster. Astronomers and Astro-Physicists just do not understand the shape or the rotation of Spiral Galaxies, Bared Spirals, or any Galaxy for that mater!!! Period!!! There is nothing in the "Theories of Relativity" or" Newtonian Mechanics" that can explain the bizarre motion or the rotation of the Galaxies. Because of this how could the mathematical model of galactic rotation realistically explain the formation of the ices ages. The truth of the matter is, it doesn't. What needs to be done is look around and see what is actually happening right here on earth to our atmosphere. The mixture of the atmospheric gasses predominately the carbon gasses is what promotes the ice ages as humanity is going to learn and they may learn it the hard way.

There is a great deal of evidence for the last ice age. This evidence is almost every where you look. There is a place in central Washington State called Dry Falls State Park. Here there is striking evidence of the last great ice age. The cliffs that the water from the Glacial run off came over, and the lake bed formed at the bottom of the falls with water still in it, are at the bottom of what once was a great cataract. There is a string of lakes that run down to the south from Dry Falls state park ending with a lake called soap lake. This is all that is left of the great run off from the melting of the Glaciers in that region

The great lakes in the eastern United States that boarder with Canada were formed by layers of ice that were almost a mile high. That in itself is truly amazing. All of North America, Europe, and Asia were covered by great ice sheets. Huge layers of ice reached clear down south as far as Arizona and New Mexico in North America. Europe and Asia also were covered with ice almost a mile thick clear down to the Alps and the Himalayas. The lower parts of south America and Africa were also covered by tremendous ice sheets as was part of Australia and all of Antarctica.

So what's so terrible about ice and Glaciers? look at what a Glacier does. A Glacier is a killer of both plant and animal life. A Glacier is a dead zone for life of any kind. There are no plants or trees that can or will grow on a Glacier. Part of the reason is that a Glacier is constantly moving. Its basic that plants and trees have to have nutrients to live and grow on and THERE ARE SIMPLY NO NUTRIENTS IN THE ICE OF A GLACIER FOR A PLANT OR A TREE TO GROW ON. Any plant or tree that started to grow under or in front of a Glacier would be destroyed by the Glaciers movement.

There are no animals that can or will live on a Glacier for any extended period of time. This is for the same reasons. The Glacier is constantly moving and is to dangerous too live on. Besides this, animals have to have plants as food to live on and there are no plants or trees that can grow on a Glacier any-way. We can cross Glaciers although they are dan-gerous to cross because of the cracks and crevasse, but we cannot live on a Glacier for any length of time. Any plants, trees, roots or seeds, in the path of a Glacier are destroyed. They are destroyed because the movement of a Glacier will grind them up an kill them. The ponderous weight of a Glacier and its movement will grind up boulders into sand, it will grind up rocks into gravel, and any plant life in its path into dust. Glaciers are killers of both types of life, plant and animal. ANY LIVING THING IN-CLUDING PLANTS, TREES, ROOTS, POLLEN, SPERM, AND LIVING CELLS, ARE GROUND TO PRIMORDIAL DUST UNDER A GLACIER.

THE SHEER CRUSHING WEIGHT OF A MOVING GLACIER A MILE HIGH IS A GRIM AND TERRIBLE DESTRUCTIVE FORCE. GLACIERS KILL EVERYTHING IN THERE PATH INCLUDING THE ABILITY OF BOTH PLANT AND ANIMAL LIFE TO REPRODUCE. THEY NOT ONLY DESTROY THE SEEDS BY GRINDING THEM UP THAT START NEW LIFE, THEY ALSO DESTROY SOME FORMS OF BACTERIA AS WELL BY THE SAME PROCESS. Just look at the terrain in Greenland and Antarctica where the great ice packs are. There are no plants or trees that grow in either place. All you see in these places is sterile lifeless earth, sand, gravel, clay. and boulders that are moved by the glaciers. All life that manages to survive in either place lives on life that grows in the sea and not on the land.

The great crushing weight of Glaciers has another sinister aspect to them that is not so obvious. Our earth is a tiny little ball of very hot plastic material with a semi-solid iron, nickel, sulfur core that is rotating independent of the exterior rotation of the planet. The continental and oceanic plates literally float on the surface of the very hot fluid or semi fluid interior material. Scientist have probed the depths of the earth with sound and ultra sonic sound waves and have a very good idea of just what the internal structure of our earth may be like. They can also measure very slight movements of any of the plates both continental and oceanic with great accuracy.

Some time ago during a major holiday when a large segment of the population in central and southern California made a large exodus to another region of the country, there was a shift in the continental plates that was actually measured. This was brought about by just the weight of many people and their vehicles all moving from one region to another. When the holiday ended and the people returned with there vehicles to their homes, a second shift in the continental plate was measured as it returned to its original position.

Recently there have been a series of earth quake tremors in Yellowstone National park and just today Jan. 9 2009 there have been reports of many more earth quake tremors at Yellowstone Park. Also there was the major earth quake that took place in China just before the Olympic games. There are earth quakes taking place all over the world everyday all of the time. These are from the movement of the fault lines where the continental and oceanic plates come together. What will happen to the earths crust when large amounts of ice are removed as is happening in Greenland? Over a trillion tons of fresh water ice was lost from Greenland alone in the year 2008. Just the movement of people and there vehicles has been measured and moved the continental plates here in the united states. What future effect will there be from the trillion tons of fresh water ice that was lost from Greenland and the ice is still melting?

When this ice is removed from the continental plate or plates in Greenland there will be a vertical motion of the plates as they rise. They will raise as the tremendous weight of the ices is removed from the plates. The tremors in Yellowstone park and the massive earth quake in China could be an ominous warning of what may come in the near future and yes the events could possibly be related. How could this be? Some layers of material in the interior of the earth are a highly fluid plastic type substance. Look at the magma that flows from the volcanoes in Hawaii. It is very fluid until it starts to cool and hardens into rock. What happens in a hydraulic system when you move a fluid in one direction? All of the fluid in the system moves and puts tremendous pressure on the whole system. Who really knows how the hot fluid or if you will semi-fluid interior of the earth will react to the changing pressure as the weight of trillions of tons of ice is removed from the continental plates in the region as it melts.

Yellowstone National Park has been identified as what has been termed a super volcano but what triggers the eruption of a super volcano? No one knows. There are good reliable theories that give a logical explanation as to what has triggered the eruption of volcanoes all over the world. There are also the fault line stress factor that explains massive earth quakes like the great one that happened to San Francisco, April eighteen 1906 . But no one knows what triggers the eruption of a super volcano. There never has been an eruption of a super volcano in the recorded and unrecorded history of the human race to find out. Yet the evidence is there that these catastrophic events have taken place as it is at Yellow Stone National Park .

The Volcanic eruptions that take place in Hawaii are believed to be from a vent that is under the pacific plate. As the Pacific plate has moved westward, Magma from the eruptions have produced the string of islands named the Hawaiian chain. The west coast of north and south America have a series of great mountain ranges that run from Alaska in the north, down the west coast of North America through central America and down the west coast of South America to cape Horn. These great mountain ranges are sprinkled with Volcanoes as is the entire Pacific Rim. The Atlantic Ocean has a trench about mid way between Europe and North America that runs from the North Atlantic south between Africa and South America in to the southern Ocean. This same trench then travels across the oceans of the world and is about forty thousand miles long. This to is sprinkled the entire length with under sea Volcanoes. As was presented earlier, it is believed that the North and South American continents are moving west ward and the leading edge of these plates are riding up over the pacific plate which is diving down under both of the American plates. This is what formed the tremendous mountain ranges that run from Alaska in the north to cape horn in the south. The Atlantic plates both north and south are spreading apart as the North and South American plates drift westward. It is believed that the westward motion of the Americas has produced all of these Volcanoes. All of these volcanoes except the Hawaiian chain are fault line volcanoes that take place where the Continental and Oceanic plates of the earths crust come together. But again this does not give an explanation as to what the mechanisms are that triggers the eruption of a super Volcano.

Now instead of removing weight by the ice melting, the ice builds up the massive mile high sheets of ice that covered most of the continents of the world during the last ice age. All the continental plates of the earth were covered with millions of cubic miles of ice. The sheer weight of six to eight million cubic miles of ice on the north American continental plate alone is mind boggling. The motion of the plates as they are now with out the tremendous ice sheets found during an ice age can and will trigger massive earths quakes and volcanic eruptions. But these are not super volcanoes. Much of California right now is on the edge of the San Andréa's fault line waiting for the big one to happen but again these are fault line Quakes. With the titanic mile high ice sheets pressing on the continental plates it isn't inconceivable that there would be Mega earth quakes and the eruption of Mega or super volcanoes like Yellowstone National Park.

Global Warming then is much more far reaching than just a change in the climate of planet earth. Global warming can and will change the entire planetary dynamics of our world from not only Glacial ice, but to earth quakes, volcanoes, and super volcanoes. The eruption of Yellow Stone National park as a super volcano would be a terrible crippling blow not only to the united states of America but to Canada as well that would take a long time to recover from. It would lay waste to most of the central portion of the Nation from way up beyond the Canadian boarder to Mexico, and from the Rocky Mountains in the west to Chicago and the Mississippi river in the east. The mid west of the united states as the so called bread basket of the world would be useless for decades to come and maybe longer.

The real cause of global warming then is an in-equality between the two life forms as Plant life on one hand, and Animal life and combustion on the other. This equality can lead to an in-equality between the two and controls the amount of the CARBON GASES in the atmosphere because of this in-equality. There always is plenty of oxygen as oxygen makes up about twenty one percent of the total atmosphere. It is the less than one half of one percent of the atmosphere that is the Carbon gasses that causes the trouble. When this tiny fraction as that part of the atmosphere that is the carbon gasses increases, it retains heat that builds up at the surface and in the lower atmosphere. This is manifested as an overpowering of one life form over the other and the gasses that make up the atmosphere.

When large numbers and tremendous herds of animal life eat the vegetation life to near extinction, and great out of control forest fires burn and destroy more vegetation over a prolonged period of time, the equilibrium between the two life forms will be up set. Animals as they eat the vegetation, breath in the atmosphere. When they breath in the atmosphere they take out oxygen and breath out carbon dioxide. Plants do just the opposite in that they metabolize Carbon Dioxide and excrete Oxygen as a wastes product. If there is a balance between the two, both will flourish. In other word, plants have to have a place where they get enough sun light, good soil and water to grow and reproduce faster than the animals can eat them. Not only this, but the plants have to reproduce enough plants so that their metabolism will take the carbon dioxide from the atmosphere that animal life it putting there. They then excrete oxygen that will maintain the balance to the atmosphere as it is. Animals in turn metabolize the oxygen excreted by the plants and excrete the carbon dioxide for more plants to grow on. The only thing wrong with this is that the animals are eating and destroying the only mechanism that takes the CARBON DIOXIDE out of the atmosphere and puts OXYGEN back into it. .(VEGETATION).

Look at the great herds of buffalo that roamed the great plains of North America before they were hunted to near extinction. Millions of them grazed on the grasslands of the mid west. You only have to look at a buffalo and it is apparent that these creatures had tremendous lungs that were capable of taking in huge amounts of the atmosphere, removing the oxygen and exhaling CARBON DIOXIDE. During the period of time that the Buffalo herds existed on the great plains of North America they put millions of tons of CARBON DIOXIDE into the atmosphere by there breathing alone. Not only did they excrete millions of tons of carbon dioxide by there breathing, they trampled and ate the very vegetation that is the only entity that removes the carbon gasses from the atmosphere. Look at the tremendous herds of animal life on the planes of the African continent right now that are still there. All over the earth there are countless herds of animals both domestic and in the wild. All of these creatures eat the vegetation including the six and one half billion creatures of the human race that includes you and me. VEGETATION AND VEGETATION ALONE ABSORBS CARBON DIOXIDE AND EXCRETES THE LIFE GIVING OXYGEN BACK INTO THE ATMOSPHERE.

Again, lightning strikes and starts forest and brush fires that burn uncontrolled, killing the vegetation placing many more millions of tons of CARBON MONOXIDE back in to the atmosphere. WHEN YOU DESTROY THE VEGETATION EITHER BY EATING IT OR BY FOREST FIRES BURNING IT, YOU ARE DESTROYING THE ONLY LIFE GIVING MECHANISM THAT TAKES THE CARBON GASSES OUT OF THE ATMOSPHERE AND REPLACES IT WITH THE LIFE GIVING OXYGEN. I CAN ONLY HOPE THAT VEGETARIANS AS WELL A MEAT EATERS FULLY UNDERSTAND THE ENORMITY AND MEANING OF THAT SIMPLE LIFE AND DEATH STATEMENT.

Once enough of the carbon gases are in the atmosphere it will be to late. Once the accelerated evaporative process increases the moisture in the atmosphere, it will accelerate the process of precipitation. Precipitation will be as rain or snow and ice. During an ice age, Glaciers covered most of the land masses of the world which are in the northern hemisphere. How could you live on land that had ice on it a mile thick? Eskimos lived on the ice in igloos, but they lived off animals that lived in the sea. There is no sea under a Glacier that is almost a miles high. The answer to this then is that animal life can not live on a Glacier of any size for any length of time. How could you grow any of the plants that you need as food to survive? Again you can not. The basic fundament needs of a human being would all be destroyed by the presence of the killer ice.

This has started right now. The summers in both hemispheres are getting hotter, dryer, and the winters more severe and cold. Not especially so much colder but precipitation will increase dramatically with greater accumulations of cloud cover and fog banks that increase more ice and snow. This increased cloud cover and denser fog banks will also shield the surface of the earth from the rays of the sun. When this becomes extreme enough, the Glaciers will start to form. Again the winters will get longer with more ice and snow storms and the summers will get much hotter and shorter. The city of Spokane in eastern Washington State has had over five feet of snow in a very short period of time and this is jut the first part of winter of two thousand nine.

CHAPTER – 4
THE EXTINCTION OF
THE DINOSAURS

The extinction of the dinosaurs has been a great mystery to science. Almost over night in geological time these great creatures suddenly ceased to exist. Dinosaurs and the vegetation that evolved with them were the dominate life form on our earth for almost two hundred and fifty million years. Dinosaurs did not evolve intellectually, instead they evolved in bulk and size. They just got bigger and bigger and bigger. Some of the vegetation eaters reached incredible lengths of from eighty to one hundred and twenty five feet. Some of these great creatures had a brain that was no bigger than a walnut in size yet many of them reached a fantastic weights of from fifty to eighty metric tons. Their huge legs consisted of four powerful pillars of flesh and bone that had to support this tremendous weight.

The great vegetation eaters were preyed upon by the great tyrant lizards. The infamous Tyrannosaurs Rex was the final great predator that preyed upon these creatures before they all became extinct. This animal reached lengths of from thirty five to forty feet and heights of eighteen to twenty feet. They walked, hopped, or jumped, (no one knows for sure) on their two powerful hind legs that supported there great upper bulk with a steam shovel sized head and mouth. Their cavernous mouth was lined with dagger like teeth and this was counter balanced by a huge powerful tail. Several of these creatures would corner one of the great vegetation eaters and literally eat the hapless creature alive. It sounds gory and brutal to us but there would have been very little if any pain endured by the great vegetation eaters as they died.

In order to experience pain, there has to be mental awarness to the pain giving activity. A human being can go into surgery and under go tremendous trauma. A surgeon can saw open your breast bone exposing your heart and then cut into this vital organ making repairs that will save your life and you are totally unaware of the activity. You do not experience any pain until you regain consciousness and then you will be given medication that softens your pain until you heal. If one of the great predators snaped the head off one of the vegetation eater then what brain there was would not be there to experience any pain anyway. It is daubtfull that a brain the size of a walnt would have the capability of experiencing any pain as we know it.

Predators right today imobilize their prey not because they are compassonate, but because they do not want their victim to escape. A lioness after the prey is dragged down will clamp the mouth of a wildebeest or buffalo shut suffocating the animal killing it before the pride feeds. It isn't inconceivable that tyrannosaurs rex simply bit the head off of its monstrous victims immobilizing the creature before the living animal was consumed.

The matabolism of these creatures had to of been tremendous just because of the sheer bulk of their size. The rib cage of a tyrannosaurs rex is huge by any standards and the rib cage of the great vegetation eaters is far larger that that. Just how many of these monstrous creatures roamed the surface of the earth before their extinction is unknown but there must have been many to say the least. Research has shown that the atmosphere of this period contained much more carbon dioxide than it does today. The high content of Carbon dioxide would have promoted the growth of vegetation during this period, providing a plentiful food supply for the great vegetation eaters. This in turn provided a plentiful food supply for the great predators that fed on the vegetation eaters.

The remnants of a huge crater found on the Yucatan peninsula in Mexico has been touted as the catalyst that brought about the extinction of these great creatures. It is believed that a planetoid or comet impacted the earth at this point causing world wide damage to the atmosphere that destroyed these huge animals almost over night. There are some calculations that put the extinction at less than a year. Much of this was by the raging fires that were ignited by the event destroying the food supply for the vegetation eaters. When the vegetation eaters starved because of a lack of food then this in turned also destroyed the food supply for the great predators. There is another group of people that dispute the impact of the celestial object as the catalist that brought about the extinction. This minority believes that there were other elements that were involved in the termination of the animals of this period.

When the earth impacted the fifth planet from the sun and the single mega continent formed that has already been presented, any and all life forms that had arisen from the primordial dust were either in the sea or on the single great continent. No one knows where or how life formed or started for that mater. The technology of mankind has not been able to duplicate the formation of life and maybe they never will. There are two life forms here on the earth that are based on the metabolism of two gasses, the metabolism of oxygen in animal life, and the metabolism if carbon dioxide in plant life. Each mirrors the other in the metabolism of life.

The change in atmospheric gasses has been shown to be dependent on the metabolism of both life forms and fire or combustion. Forest fires which is the combustion of vegetation and the metabolism of animal life increases the carbon content of the atmosphere. An increase in the growth of plant life and a decrees in the growth of animal life decreases the Carbon content of the amosphere. The carbon gasses are heavier than the atmosphere, so if left undisturbed will settle down towards the ground and the lower atmosphere in most cases.

Reptilian life that evolved during the evolution of the dinosaurs did not evolve intellectually, but they did evolve to greater and greater bulk. They just simply got bigger and bigger. With the great Herculean mass of their existence, they had a bigger effect on the carbon content of the atmosphere by their metabolism. These huge creatures placed tremendous amounts of Carbon Dioxide into the atmosphere by there respiration alone. It would be hard to imagine the amount of dung produced by the sheer size of these animals. Dung produces menthane which is another carbon gas although to a lesser degree.

If the great revolving storms that we have today did not form as often or maybe not at all then the carbon gasses created by these creatures would have stayed close to the ground and in the lower atmosphere which is the Troposphere. This would have accelerated the growth of vegetation greatly during this period. The tenure of the dinosaurs came to an end about sixty five million years ago! The great mega continent formed a tectonic plate deep split or crack from north to south that separated the North and south American continents from Europe, Asia and Africa! There is evidence and opinions that that this happened about one hundred fifty to two hundred million years ago.

There is also evidence and opinions that the atlantic ocean opened and closed more than one time. Just how great was the opening and closing of the atlantic ocean is an unknown. No one knows if this event opened the waters all the way to what it is today or maybe just half, or maybe even just a quarter. The point is, if this opening and closing of the atlantic ocean was maybe just a quarter of what it is now then the great revolving storms we contend with today would not have formed as often as they do now or maybe even not at all. These great revolving storms have to form over open warm water and not land. They have to have a long fetch of warm open water to develop and the closing of the Atlantic ocean would have eliminated this. These great revolving storms are great mixers of the atmospheric gasses of the atmosphere. With out them the heavier carbon gasses produced by these huge animals and the combustion that produced more carbon gasses by uncontrolled forest fires would have remain closer to the ground and in the lower atmosphere. There would have been nothing to stop forest fires during this period but the great concentrations of the carbon gasses close to the ground and in the lower atmosphere. These concentrations of carbon gasses would have been a deterrent to the spread of any fires during this period. We use Carbon Dioxide in fire extinguishers to smother and put out fires right to day.

When the object impacted the Yucatan peninsula in Mexico, it mixed the atmospheric gasses by the impact. The heavier carbon gasses in the lower atmosphere were blown away and mixed with the rest of the atmosphere by the traumatic happenings of the event. The vegetation, plant life, etc that had evolved with the eveloutation of the great beasts had to have a rich mixture of the carbon gasses to survive that was provided by these huge animals. They were suddenly stripped of the protective shield of carbon gasses that they depended upon for their warmth and their metabolism. When this was removed almost over night, they just died pure and simple. Millions of square miles of dead and dying vegetation would have made tremendous reservoirs of fuel for great forest fires all over the world. All that was needed was a single bolt of lightening to start it. With out the vegetation for the great plant eaters, they died and of course with out the great plant eaters as a source of food the great predators also perished. After the object impacted the earth, the dead and dying vegetation would have burned uncontrolled in every continent on earth.

CHAPTER – 5
THE GREAT AND TERRIBLE EXTINCTION

What is habitat and what is needed to sustain life? The definition given in the American dictionary is, "The area or environment in which an organism or ecological community normally lives or occurs." "The place a person or being is normally found." So the planet earth is the area in which the organisms or ecological communities of all life on earth exists and this includes the habitat of mankind. Habitat is your house or apartment, a hole in the ground for a squirrel, or chipmunk. When you destroy habitat then the creatures that occupied the habitat must adapt to a new habitat or perish.

If ten people were placed in an air tight room, with no ventilation, all ten would die in a short time. Their metabolism would deplete the oxygen in the room replacing it with carbon dioxide and they would all suffocate. If there were enough plants placed in the same room with the ten people, the plants would remove the carbon dioxide from the air and the people would not die. The plants not only remove the carbon dioxide, they also excrete oxygen back into the air that is vital to the survival of the people.

What most people do not realize is that the planet earth is a closed air tight room with all life both plant and animal in the same room as the habitat of life. The atmosphere that we breath and the gasses that make up the atmosphere are central to the earth as a closed room that is the habitat of all living things. There has to be enough living plant life to keep the carbon gasses to a low enough level so that they do not effect the atmosphere raising the temperature of the earth. The areas where plant life will grow are limited to the tropics and the temperate regions yet we and much of animal life roam all over the earth.

Not if, but when the great Glaciers start to form, the habitat of all animal life on all the continents of the earth including the habitat of humanity will change. Under such conditions there would no doubt be a mass migration of animals both domestic and wild to escape the encroaching ice. Animal life has to consume plant life to live on and survive. Or! Animal life has to consume Animal life that lives on plant life to survive. Either way Animal life has to have plant life to live on and survive and there is no plant life on or under a Glacier. Because of this there would be a mass migration of animal life.

There would also be a mass migration of the human race for the same reason although this would be much more complicated. We have countries and different nations to contend with. Borders would have to be crossed, and different cultures with different societies that would have to mingle. The entire social structure of humanity would disintegrate and start to break down.

What would the people and the government of the United States of America do if all the Canadians from Canada tried to migrate into the United States at one time because of the encroaching ice? What would the Mexican government and the Mexican people do if all the people of the United States and of Canada combined tried to migrate into Mexico at the same time for the same reason? The same thing holds true for Europe, Asia, south American and the African continent. What would the governments and the people of Indochina, Thailand, Cambodia, India, and the African continent, do if all the people of the northern countries say all of Europe, the Soviet Union and the Chinese all tried to migrate into their countries? There would be mass chaos on a grandiose scale there would! Officials at the borders would try to stop them at first until the massive crowds pushed and shoved there way past and then chaotic pandemonium would ensue.

All over the world humanity in there desperation for a place to stay and land to live on would try to flee south in the Northern Hemisphere and north in the Southern Hemisphere to escape the building ice sheets. Look at some of the states in the south western United States right today. Here in Arizona you can see what it was like during the last ice age. There are the flat tops of mountains every where that have been flattened by the wave action of monstrous lakes. They Indians call them Mesas. They look like somebody came along and cut the tops of them off with a knife. There are also the water marks as a dark line on some of the mountains in many of the lower western states that are visible even today. What are desert areas now, were once huge lakes all over the region. The evidence is there from the great salt lake and the great salt lake salt flats to the now dry, desert, old lake bottoms all through Nevada, southern California, some portions of Utah, Arizona and New Mexico but they weren't very pretty then.

The run off of water forming the titanic rivers from Glaciers towering almost a mile high stretching clear across the United states would be ice cold and contain chunks of ice and yes even ice bergs that periodically break off from the front of the Glaciers. You see this happening in Alaska right today where the Glaciers meet the sea. The increased atmospheric temperatures of Global warming fuels grossly accelerated evaporation from the seas and everything else. This evaporation includes moisture from plant life much of what dies from the increased heat primarily in the tropical regions.. The increased heat not only accelerates the evaporation of plant life but animal life that includes all the six and one half billion people that make up the entire human race. The accelerated evaporative process on a Global scale dramatically accelerates the precipitation process with increased cloud cover in the northern and southern regions of the world. This in turn accelerates bigger more powerful wetter storms and longer harsher winters and shorter hotter summers in both hemispheres.

This is happening right now in the winter of two thousand and eight and two thousand and nine. As I write this there is a blast of cold artic air with record breaking cold temperatures covering most of the eastern part of the United States with heavy snow and SHEETS OF ICE. Many of the eastern states are paralyzed and with out power by the damage caused because of the ice. Ice forms when the surface of the earth is cold, well below freezing and then it rains. The water in the form of rain from a warm atmosphere when it hit's the cold earth freezes it into ice. It not only damages power lines, it damages and kills trees and brush by breaking the limbs off and even uprooting them. This curtails, hampers, and destroys the ability of vegetation to remove the carbon gasses from the atmosphere. Vegetation removing the carbon gasses from the atmosphere is as important as curtailing the release of the carbon gasses in the first place.

Fires of any kind put carbon gas right back into the atmosphere. The burning of vegetation does more than that. The burning of vegetation not only puts carbon gas back into the atmosphere, it kill the vegetation that is the only way of removing the carbon gas in the first place. Just this last summer in two thousand and eight there were as many as eighteen hundred forest fires burning out of control all over California and in most of the western states. How many more all over the world were burning that never were reported is anybody's guess. As the summers get hotter, there will be droughts that will turn more regions into deserts destroying vegetation. Hotter summers destroy vegetation in the summer. Heavy snow and ice storms destroy vegetation in the winters. Again ,Vegetation is the only mechanism that captures the carbon gasses and removes them from the atmosphere.

There will be two areas fronting both of the Glacial systems. One of these is in the northern hemisphere as the northern Glacial systems and the other is in the southern hemisphere as the southern Glacial systems. Both of these systems will face the torrid hot tropical regions straddling an equatorial belt that circles the entire globe. Where the frontal Glacial systems meet the hot tropical regions of the equatorial belt, they rapidly melt, forming the tremendous run off rivers and lakes that are fed by run off of both systems. This then is the accelerated, evaporative, precipitative structure of Glacial ice and their weather systems. These weather systems are what builds the tremendous ice sheets and the accelerated Glacial run off that forms the Mega lakes and titanic rivers from the melting ice.

The Mississippi river basin in the central united states and the Columbia river in Washington state are what is left as remnants of the great rivers of the last ice age in North America. All of the lower western part of the United States and most of Mexico were flooded with huge ice cold flowing lakes laced with pieces of ice and small ice bergs. Part of the Glacial system formed a drainage between the Cascades and the sierra Nevada mountain ranges in the west and the Rocky Mountains to the east that emptied into the sea of Cortes in the south. The only dry land available were the tips of mountainous rocky outcrops that you can still see today if you look. The Rocky mountains above ten thousand feet in the west and the Appalachian mountains in the east were all the land that projected above the ice of the Glaciers. Evan so, they to would have been under monstrous ice sheets that formed on their slopes.

There are six and one half billion people on the earth right now, and if I should be fortunate enough to live to be one hundred years old it will be the year 2030. It is estimated that by the year 2030 that the world population will increase to over EIGHT BILLION PEOPLE. Where would all of those billions and billions of people of the earth, and all the animals both domestic and wild go? Into the sea? Into the atmosphere? Into the earth? NO?!! The sheer horror of the situation is, that once the Glaciers start to form and the monstrous lakes and rivers created by the run off from the Glaciers form, there would be no place for animal and human life to go and live on any where in North America. Not only would most of the North America continent be affected, but all of Europe, England, most of Asia, the southern portion of south America, the southern portion of Africa, most of Australia and all of Antarctica would be covered with building ice sheets. Once the Glaciers start to build and temperatures fall below freezing, no plants, trees or vegetation of any kind can grow anywhere the Glaciers are. The only place in North America would be Mexico and central America and they to would be flooded by the monstrous flowing lakes created by the Glacial run off. THE ENTIRE NORTH AMERICAN CONTINENT WOULD BE UNINHABITABLE. JUST TO CONTEMPLATE SUCH A DISASTROUS TURN OF EVENTS BROUGHT ABOUT BY GLOBAL WARMING IS UNTHINKABLE. The same destructive forces that brought about the untenable living conditions in North America would of course be prevalent on all of the continents of the world.

Once enough of the carbon gasses are in the atmosphere and the ice starts to form, the ice destroys the only mechanism capable of removing the carbon gasses from the atmosphere! Vegetation! There is very little if any vegetation that will grow when temperatures get below the freezing point of water let alone smothered under building sheets of ice. Much vegetation dies and leaves seeds to reproduce when it gets below freezing. The first hard frost kills the leaves on the trees and then they loose them and go into a dormant state waiting for better weather to come, but what if it does not come, then the trees will all eventually die.

When conditions becomes extreme enough, this then is the setting for the mass extinction of most of the life on the continents of earth. Not only the entire human race and all animal life but the vegetation as well. The Glaciers are killers. Any life form either human or animal that tried to stay on the Glaciers would either starve to death from lack of food, be frozen by the cold, or be killed by the moving ice. Those that tried to live in the south western part of the North American continent would also either starve to death or be drowned in the frigid mammoth flowing lakes that covered most of the region. The only dry land available would be the rocky mountainous out crops that extended above the flood waters in the lower western part of the united states. You can still see many of these mountainous rocky out crops in some of the south western states today. People marooned on a rocky out crop would die of starvation or drown in the swift moving ice cold currents trying to escape to another rocky out crop. If they did escape to another rocky out crop they would just starve to death on that outcrop.

The deep south of the eastern United States would not be any better off than the south western portion as most of it would be covered by a mammoth ice cold lake hundreds of miles wide. They think they have a five hundred year flood in the Mississippi valley now, wait till the Glaciers and the run of from the glaciers form. There will be a lake where the Mississippi river is now, from New Orleans west to the rocky mountains and to the Appalachian mountains in the east. Chicago and the great lakes will be under a towering Glacier of ice and snow almost a mile high that stretches clear across the entire North American continent down to Arizona and new Mexico. It has happened before and believe me it is starting to happen again right now. It is happening many times faster now than it has ever happened at any time in the past history of the world because of what we are doing. (WHAT WE ARE DOING NOW, IS THE THOUGHT LESS, RECKLESS, BURNING OF FOSSIL FUELS)

The tropical regions of the world would be almost as bad if not worse. The build up of heat in the tropical regions after the land dried out due to the heating effect of global warming would dry out huge areas of Africa, South America, and Asia into vast arid desert regions. Death Valley in the south western part of the United States of America has recorded a record temperature of one hundred thirty four degrees in the summer of nineteen thirteen. There they record temperatures every summer well over one hundred and twenty five degrees. How could plant life and animal life live in temperatures of say one hundred and forty degrees, one hundred and fifty, one hundred and sixty degrees? You cook food to a temperature of one hundred and sixty degrees to kill E coli and when you do, you kill most of the bacteria in the food. A temperature of one hundred sixty degrees will cook your flesh and kill you.

You try to live in an area as habitat that has temperatures in excess of one hundred and forty degrees and you and anything that you tried to raise for food would die. The internal temperature of a human being is very narrow at ninety eight point six degrees. If this internal core temperatures is raised as little as eight degrees you are very near death. We sweat so that the evaporation of water from our skin will carry the internal heat from our bodies maintaining our uniform core temperature. When you sweat you have to take in large amounts of fresh water to keep replacing the water you are sweating out. At much lower temperatures than ninety eight point six degrees you start sweating to cool off and maintain your core temperature. If your core temperature is lowered as little as ten degree you are suffering from hypothermia and you again are near death. At temperatures just a little less than seventy degrees, we start looking for an out side source of heat to keep warm and maintain our body core temperature.

As the temperatures increase, rain when it falls in the tropical regions of the world will not reach the ground. It will evaporate before it ever gets there. This means that there would be very little fresh water available in the central latitudes on the continents if any. The higher temperatures because of the increased heat will dry out everything on the land in the tropics and kill most vegetation forming vast deserts. Where the Amazon basin is now in South America, and the central regions straddling the equator of Africa and Asia, would be a raging inferno that would look like a burned out region of hell as the heat builds. Temperatures in these regions could reach as much as maybe one hundred forty degree to one hundred fifty degrees. Dry hot thunderstorms will discharge lightning strikes that start forest fires that burn uncontrolled compounding the situation but it is to hot for it to rain.

That happens a lot here in Arizona. We are in the weather rain shadow of the Sierra Nevada and other mountain ranges to the west of us in southern California. You can see the rain falling from the upper atmosphere in the clouds, but it never reaches the ground. When it falls to the lower level where the heat is, it evaporates right back into high humidity. You can also see clouds disappear right before your eyes. Here in the desert south west where the humidity is low, whole clouds systems that blow in during the daytime, evaporate into high humidity in the late after noon and evening. The sky's then becomes clear at night.

All life, both animal and human in central South America, central Africa, and all land that straddles the equator clear around the globe on both sides, would try to migrate to both the north and the south. They would move trying to escape the blistering heat caused by the raging inferno of the tropics They would not only flee from the fires, but the infernal concentrations of building heat on the land itself. The very ground itself at a hundred and forty to a hundred and fifty degrees would be to hot to walk on. The heat will continue to build every hour, every day, and every year because of the carbon gases in the upper atmosphere with no way to take them out. Jet planes, Hurricanes, Cyclones, Monsoons, Tornadoes, all of these entities mix the carbon gasses high into the upper atmosphere where there is no vegetation to recapture them. The only way to remove the carbon gasses from the atmosphere is with plant life and vegetation will not grow on Glaciers a mile high and certainly not in the upper atmosphere where the so called modern jet engine is depositing them. The carbon gases trap the heat and every day the sun shines on the earth it gets just a little hotter and a little hotter.

Alas, as Animal and Human life tried to escape the terrible heat of the tropical regions they would come face to face with the mile high towering Glaciers to the north and to the south and the great ice cold flowing rivers and lakes that are the run off from the Glaciers. The life giving vegetation of the earth now is almost completely gone. It has been , destroyed by the Glaciers, the terrible searing heat of the tropics, and the raging cold flowing lakes and rivers that are the run off of the monstrous Glaciers. What was left of the great forest of North America, Europe, and Asia after mankind had logged them to near extinction would be covered by snow packs ALMOST A MILE THICK These snow packs are what compress's and forms the great killer Glaciers that covers most of the land. They literally smother and grind what vegetation is left on the continents into oblivion. The terrible heat and ragging fires in the super heated tropical regions will destroy what vegetation is there and make the tropical regions un-inhabitable.

All over the entire world there would be a mass migration and then crowding of life together both animal and human with no food, shelter, or sanitation for any. All life both animal and human in the region of the Glaciers would try to migrate into the tropical zones just to escape the advancing killer ice. All life both animal and human in the tropical regions would try to migrate to the regions of the Glaciers both north and south in order to escape the searing dry heat of the tropics.

There would be a mass crowding of animal and human life in the two regions. The two regions are between the destruction of the Glaciers from the north and south and the inferno of the tropics in between. There would be no place for all of the hapless creatures that have inhabited the entire earth including what is left of the human race to go. All would effectively be trapped. Much of the human race would have already been killed by being suffocated to a slow death right in there homes under the relentless marathon snow and ice storms that built up and became the Glaciers. The toll in human suffering would be staggering. Death will be everywhere. Plant life cannot grow because of the advancing snow packs that form the ice sheets that extend from the north and south and from the terrible searing heat in the tropical regions in between. All the plant life where the Glaciers are would be destroyed by the Glaciers All plant life in the tropics would be destroyed by the searing concentrations of heat. This means that all animal life including all members of the entire human race are faced with MASS STARVATION.

When there are no plants to eat, Animals will eat Animals!! Animals will eat Humans!! Humans will eat Animals!! and HUMANS WILL EAT HUMANS!! just trying to survive. Yes!! the sheer horror of this situation is, that when the human race is faced with mass starvation, they will turn to cannibalism. There is always that small portion of society that will commit crimes like, robbery, mayhem and murder and when faced with starvation they will turn to cannibalism. This small portion of humanity is what will start the trend toward cannibalism. The great civilized values of humanity will be reduced to the basic law of the jungle and survival, which is simply eat or be eaten. This will happen not only here in the United States of America, but all over the world. As grisly and ghoulish as this sounds, it is just exactly what will happen. Roving bands of half crazed starving teenage killers will attack other bands of people singling out women and children killing them and use them just for food. Look at the lower species of animals and what they do. Predators when they hunt look for the young, the weak, and the old. Lions, Tigers, all of the big cats, Wolves, almost all predators that hunt, do so and many of them hunt in a pack. They single out there prey, then they attack and the whole pack tears the hapless creature to pieces. The pack then fights over and feeds on the remains of the victim.

The life span of a human being very seldom exceeds one hundred years. This means that if the human race ceases to reproduce its self, with in a scant hundred years we will become extinct. What children are born into this scenario from hell are prey for the roving bands of mindless cannibalistic killers. The end of life on the continents of Planet earth now is nearing its final end.

The last vestiges of life in the Americas are in southern Mexico and central America. There are so many people trying to find some place to go, that they literally stand almost shoulder to shoulder in their pitiful stinking filthy misery and stark naked starvation. Insanity is rampart. The last and final death thro of human life on earth is in its final moments. People everywhere are crying, People are cursing, and People are praying. Those that tried to escape by going to sea would find the same conditions all over the world. When ships at sea run out of fuel, they either sink when battered by storms or end up on some beach some where to be broken up into relics of the great and terrible ice age. The only life left on the planet earth now, is in the sea.

Did you ever wonder why dolphins and sea otters that are air breathing animals returned to the sea after evolving on land? Could it be that they returned to the sea during a killer ice age when the land masses became uninhabitable for life either plant or animal. Some plant and animal life including human might manage to survive the massive destruction by escaping to the mountain tops if they weren't covered by Glacial ice. Any place on earth where the mountains projected above the major ice sheets and were not covered by ice sheets them selves might possible be a haven for life. This could be on volcanic slopes, that portion of the Himalayas in Asia and the Alps in Europe that were in a tropical climate. The entire coastal range of the Americas that stretches from Alaska in the north to the tip of cape horn in south America. In fact any of the mountains on earth that were not themselves covered with ice sheets could possibly be a haven for some life for a time if they could find or grow food.

Would the human race if faced with mass starvation turn to cannibalism? Look into the archives of the earth. There were tribes in the south sea islands that engaged in cannibalism although they had plenty to eat. Fruits, Nuts and sea Food, but no meat. When grandma or grandpa passed on they had a big celebration. They roasted the deceased grandparent and had a great feast. There rational was that by consuming their ancestors, they would always be with them. To our western way of thinking, this is sickening.

Neanderthal man existed during the last ice age. He was a short stocky brutish man like creature with a limited intelligence. It is believed that Neanderthal man never numbered much more than 15,000 strong at the peak of there existence. Neanderthal man was a hunter who hunted animals as food but they were also cannibals. Finds made by Anthropologists confirm this but why was Neanderthal man a cannibal if they could hunt game to live on?

During the last Ice age plant life of any kind was in short supply. During this period, most of the earth was covered by monstrous sheets of ice and the tropical regions were a burning dry hell that could not support much vegetation. With out plant life, animal life will perish. This would have made a shortage of game for Neanderthal man to hunt for food.

Most creatures including modern man will not turn to cannibalism unless there is no food and they are starving to death. Even then there are those that no matter what the end will be, they will not turn to this last degrading resort for survival. They will just simply starve to death. When they do they could be eaten by those that will.

Once Neanderthal man turned to cannibalism it became a way of life for them. They may have formed families or clans that roamed the country side hunting for food. True they hunted animals as food but they still engaged in cannibalism and when game was scarce, they simply hunted and killed each other as food. One clan or family would kill or capture another clan or family, eat the dead killed in battle, bury their dead and then keep the living as prisoners for a future food supply. It was a harsh and brutal way of life. It was ghoulish, grim, and disgusting by our standards but they survived. All findings made by the scientific community point to just that.

Neanderthal man became extinct and again why did this happen? Homo Sapiens which is you and me originated to the south in the dark African continent and migrated from the tropics to the northern latitudes during the end of the last Ice Age. There they encountered Neanderthal man and their cannibalistic ways. Homo Sapiens would have been the target as prey for the cannibalistic Neanderthal Man. The only problem was that

Homo Sapiens was a lot smarter than Neanderthal man so Homo Sapiens with there much greater intelligence simply hunted Neanderthal man to extinction. There is evidence for this because as Homo Sapiens moved north, villages and home sites that Neanderthal man occupied, just disappeared but Homo Sapiens survived to this day. How long would it take for life to rebuild back to what we have today if it ever could? Would it take fifty thousand years, a million, ten million. a hundred million? Who knows. What tattered remains of the human race that were left if any would live as stone age people. All of the technology that we pride our selves on today would be gone. It is heart wrenching, and sort of takes your breath away to contemplate such a fantastic terrible series of events happening. All the great cities that man kind has ever built, Chicago, New York, Detroit, Seattle, San Francisco, Moscow, Paris, London, Beijing, Calcutta all over the world, all of the great cities that the human race has created would be gone. Obliterated!!! Destroyed!!! Wiped off of the face of the earth for ever. It makes a person feel kind of melancholy, sad, and lost to contemplate such a dramatic and final end to all of humanity and our great civilizations. All the great cities that man kind has created were either incinerated by the searing heat in the tropics, ripped apart by the ice cold raging flood waters, or they were GROUND INTO OBLIVION by the shear weight and motion of the monstrous mile high killer Glaciers.

This is a grim and terrible projection of what will happen in the future if we do not control Global

Warming. It is a very real and deadly thing that has started to happen right now. It has happened to the earth before in the past with out the coal fired utilities, the oil and gas companies poring Trillions of tons of carbon gases into the atmosphere. By digging up fossil fuels and adding them to the carnage, it is doubtful if the human race will survive the next ice age at all and this ice age is going to happen tomorrow.

CHAPTER – 6
AN ETERNAL POEM WHO WILL EVER KNOW

BY

VERN G. RICKEY

WHO WILL EVER KNOW, as the last living child runs down a wash?

WHO WILL EVER KNOW, as she is over taken by the swirling waters that throw her head long into the mud?

WHO WILL EVER KNOW, and hear the child like screams, Ma-ma, Ma-ma coming from her lips as they are finally and mercifully sealed forever?

WHO WILL EVER KNOW, that her efforts are in vain as she struggles with the rising water and the mud that entombs her for all the eons.?

WHO WILL EVER KNOW, that these are the last and final words that were ever utter by the last living member of the human race?

WHO WILL EVER KNOW, who she was?

WHO WILL EVER KNOW, of the great love her mother had for her father, and the great passionate love he had for her?

WHO WILL EVER KNOW, of the great love that women have had for theirs, and the great love that theirs has had for them?

WHO WILL EVER KNOW, of the beautiful cities that were created by mankind and then were ground into oblivion by the mile high killer glaciers?

WHO WILL EVER KNOW, of the men that walked on the moon and all of the brave that died in all of the wars of mankind?

WHO WILL EVER KNOW, what the hopes and dreams were of an extinct race that perished because they DID NOT KNOW?

WHO!!! I ASK YOU, IF THERE REALLY IS SUCH AN ALL POWERFUL ENTITY

AS GOD IN EXISTENC??? WHO IN GODS ALL MIGHT NAME, COULD EVER

REALLY KNOW???

CHAPTER - 7
WHAT THIS MEANS

WHAT THIS MEANS, IS THERE IS GOING TO BE ANOTHER ICE AGE AND IT IS GOING TO HAPPEN RATHER QUICKLY. BY QUICKLY I MEAN IN TEN TO FIFTY YEARS AT THE MOST. Natural catastrophes happen rather suddenly. An Earth quake does tremendous damage and can happen in just a few seconds. Even the worst earth quakes do not last over several minutes. A tornado will touch down and just devastate a community and be gone in a few minutes. Hurricane Katrina came in and in a few hours almost completely destroyed the great sea side city of New Orleans. A tidal wave will suddenly come up out of the ocean and smash a coastal community wrecking tremendous havoc and cause great loss of life. Those mile high Glaciers of a new ice age now would be able to build up in an incredibly short period of time. How could this happen?

First lets look at a few of the facts, not opinions but facts. The ice ages are something that has actually happened. There just is to much hard evidence that has been uncovered by the scientific community that can not be disputed that bears this out. This evidence is not only here in the united states of America, but on all the continents all over the world. Using dating techniques and empirical data gleaned from geological sites on a global scale, it has been shown how the last ice age ended about ten to twelve thousand years age.

Second, Somehow all those millions of cubic miles of water from the oceans of the world moved from the sea to the land and froze there as ice for millions of years. It has also been shown how the level of the worlds oceans have dropped significantly during an ice age. Sea levels have dropped as much as four to five hundred feet. One scientist came up with a figure of four hundred and ninety feet.

Third the only way that amount of water could get from the sea to the land is through the process of evaporation and precipitation. There just is no other way for this to happen. Water does not evaporate when temperatures get below freezing it changes into a solid State which is Ice!! Water will not run up hill in either a liquid or solid state so it did not run or flow up on to the continents.

Fourth, there had to have been somewhere between thirty and fifty million cubic miles of ice on the continents of the earth during the last ice age. Think about that. Thirty to fifty million cubic miles of ice. Great sheets of ice covered most of the continents of the earth.

Fifth it takes heat to evaporate water. Because of this all the water had to be in the form of vapor when it was deposited on the land. Water can only get into a form as vapor when it is heated so the earth had to of gotten hotter in order for the water that formed the glaciers to have gotten on to the continents of the earth.

And finally sixth, what is the actual difference between winter and summer in the polar regions of the world? The only real difference is simply shade. This shade is produced by the tilt of the earth on its axis of rotation in its annual orbit around the sun. The Artic region of the earth is actually closer to the sun in winter and further in summer. Yet the tilt of the axis of rotation of the earth places the Artic region above the artic circle in the shadow or the shade of the earth itself during the winter months. Because of this even thought the earth is closer to the sun during this period, the suns rays can not reach the surface of the earth there by becoming heat and warming it. Heat rises, and if it is not constantly being replaced by the rays from the sun at the earths surface then temperatures at the surface will drop and continue to drop. This happens every night as the earth spins on its axis of rotation. Because of this simple difference, temperatures can vary between seventy to eighty degree above zero in direct sunlight in the summer months to as much as sixty, seventy, to eighty degrees below zero in the winter months. The temperature spread is much wider in the Antarctic region because the distance from the sun is reversed and this pole is farther from the sun in winter.

If large areas of the earths surface in the polar regions become covered by vast massive thick clouds and dense far reaching fog banks, then the sun light can not reach the surface of the earth for extended periods of time in these regions. The sun light can not be converted to heat there by warming these areas. The heat that is in the ground will continue to raise from the surface up into the atmosphere. When this happens the ground will get colder and colder as heat is lost up into the atmosphere freezing water and forming frost on the ground. When this cloud cover, and fog banks become massive and dense enough, they then will shade the surface of the earth year round in the polar regions. Both of the polar regions of our world will become shrouded in thick heavy cloud cover forming massive polar caps of clouds that stay year round blocking the sun light from ever reaching the ground. This will then turn the summer months into a prolonged year round winter. As the accelerated evaporation process in the tropics continues, the polar caps of cloud cover will increase and will extend farther north and south from both poles covering a large part of both the northern and southern hemispheres of the earths surface.

All it has to do is get cold enough to snow. It doesn't even have to do that. When the atmosphere is warmer and is moisture laden with excessive water, and the surface of the earth gets colder and below freezing because of no direct sunlight, when it rains the water will freeze as ice on the surface as during an ice storm. It is doing just that right now in the winter of two thousand and eight and two thousand

and nine. There have been several great ice storms in the mid west and eastern United States. These ice storms deposited only about three inches of ice and yet the east coast of the united states and everything there was a mess for almost three weeks. Transportation, and all human activity was paralyzed during this storm. Human activity did not return to normal until the clouds cleared, the SUN COULD SHINE AGAIN, and finally melt the ice. What would happen to the people there if the storm had deposited a full foot of ice. Worse yet ten feet. One hundred feet. A thousand feet. Five thousand feet. The ice in the last ice age was almost a mile high (5280 feet) in many place all over the world.

England had a record snow storm in two thousand and eight that paralyzed this country for a while. What if the snow storm there had not stopped? With a huge increase of moisture in the atmosphere, there would be an increase in the length and intensity of storms world wide. When the Glaciers start to form, they will not be re-melted starting in the higher Artic and Antarctic latitudes during the summer months. They will not be re-melted because of the increased heavy fog banks, mist and thick cloud cover covering both of the polar regions. The sun light will not be able to reach the ground because of this and the ground will become colder and the glaciers will start to build. When the sun light can not reach the ground because of the heavy cloud cover, photons of light will generate electrons in the cloud tops leading to massive charges of electrical energy in the clouds. When these charges become great enough the electrical energy is discharged as simply lighting. The massive heavy thick cloud cover then becomes huge powerful intense storms. When this starts, every year during the winter months they will build the glaciers layer after layer and grow as long as the carbon gasses are in the atmosphere. They will not melt during the summer months because of the heavy cloud cover covering the polar caps. As they build they will extend further south in the northern hemisphere and north in the southern into the temperate regions.

Hundreds to thousands of extra cubic miles of water will evaporate from the sea into the atmosphere because of the extra carbon gases in global Warming. The way things have been in the past, a storm or blizzard will only last just a few days up to a week at the most. The reason that they do not last any longer than this is because there hasn't been enough moisture in the atmosphere for a storm to lasts any longer than that. After the moisture has fallen out of the atmosphere, the storms ends. When the evaporation precipitation process is accelerated by the atmosphere becoming warmer, it will start to rain and snow over north America, Europe and Asia and in the southern hemisphere. When the rain and snow starts to come down that form the great Glaciers, it may last for weeks, months, and maybe even years. Then what could people do? At first they would try and dig out but how could they keep up when the snow and ice just kept coming and reached say ten feet. Fifty feet. five hundred feet. Again the great glaciers of the last ice age were just a littler under a mile high and they stretched clear across north America, Europe and

Asia. They extended as far south as Arizona and New Mexico in North America and south to the Alps and the Himalayas in Europe and Asia. They covered much of the south part of south America and the southern portion of Africa,. most of Australia and all of Antarctica. The great civilizations that we have to-day (billons of people) would be buried alive under the falling snow and ice storms with no chance of escaping. Even a mini ice age with glaciers a couple of hundred feet thick would be devastating.

This is a hard subject to address and the human race has been avoiding it but it is of such a vital importance that sooner of later we are going to have to come to grips with it. We simply are going to have to learn is to control our numbers. The population explosion is of a growing concern and something is going to have to be done about it. I don't mean some of inhuman atrocities that have been committed by people like Adolph Hitler. That fiend from hell murdered six million people in cold blood. Another was that dim witted imbecile of a Pole Pot. This man murdered almost a fifth of his country men just because they were intelligent. This beastly, extremely stupid brutal pheasant of a man murdered any one that had any intelligence about them at all. All this sub standard inhuman clod did with his mass murder of his country men was show the sheer rank stupidly of his inferior sub standard but sickly perverted mind. I can not find words to express my complete and utter contempt for this disgusting thing. You certainly can not call this malfunctioning atrocity a human being.

Sadam Hussein, the great murder Stalin, the list goes on and on and look at the carnage that is going on in Africa right now. The senseless murders, the gang rapes, the complete disregard for human rights of any kind. All of the middle east are fighting over Religion, land, and Oil and to what end.

The population explosion is a great and growing threat to mankind. There needs to be a meaning full direction for the human race to follow. We have to know where we are going as a species. One race or one religion cannot hold its self superior to another and yet they do. We have to recognize that plants, trees and all vegetation are the other half of life on earth and again they do not. With out the vegetation on earth as the other half of life, We will all die.

The equality and rights of every living creature on earth have to be respected and taken into consideration. This includes animals in the wild and the animals that we raise and kill for food. The Chinese were trying to address the problem but they to were almost as brutal and inhuman about it as that dim witted Pole Pot. We have to learn to control our numbers but for gods sake be humane about what we do. Life for the individual is precious. I have a small dog as a pet and his little doggy life is as precious to him as my human life is to me. The only humane way to control our numbers is with birth control. Birth control on a global scale. Why not start with voluntary vasectomies for men as one. Offer the operation strictly as a voluntary decision. Educate people to the dangers of the direction that we are going and what the complete disregard for plant life as the other life form on earth will do to our existents.

It is vital that the pope and Catholic church see the danger to humanity and change there stance on Birth control. What was good for yester year is not good for the present day world. Birth control is not meant to promote promiscuous behavior, but is a means of controlling our numbers. If mankind does not learn to control our numbers in a humane and compassionate way, then the forces of a brutal but natural system of evolution will and it will not be nice or pretty.

There will be more frequent senseless wars that kill people, more virulent diseases from over crowding that kill people, more mass starvation that kill people, less tolerance of each other and greater more atrocious acts of inhumanity committed by humanity against humanity including the most despicable forms of torture. Mankind has got to learn that this little earth is only so big and the population explosion can not continue unabated forever with out grim an terrible consequences happening to the human race.

CHAPTER - 8
WHAT CAN BE DONE

And finally what can be done? The sad part of it is, that it may already be to late. The way we are releasing and spewing Carbon gasses into the atmosphere on a global scale is frightening and the greatest folly ever committed by the human race. A single coal fired power plant puts one hundred million tons of carbon gasses in to the atmosphere every year and there are thousands of power plants just here in the United States of America alone. China, Russia, India, South Africa all the developing nations world wide are burning coal at an ever increasing rate and they are releasing fantastic amounts of carbon gasses into the atmosphere that we have no control over.

Some of the things not to do, is to start killing all the animals on earth, both in the wild, and domestic. It is not going to do any good by killing anything. That not only goes for animals, but people as well. There are already to many people being killed in senseless useless wars around the world. Killing anything would be counter productive to what the real problem is. The real problem is that there is just not enough vegetation on the earth that takes the carbon gasses out of the atmosphere to support the existing animal life that is here. This includes all six and one half billion members of the entire human race and that number is steadily growing. It is not only the six and one half billion people of the earth, it is the activities that they engage in utilizing the combustion of fuels of any kind that produces the carbon gasses.

Again remember, that it is only about one half of one percent of the total atmospheric gasses that are the Carbon gasses and the people that make up the human race are the ones that are changing this at a alarming rate. Our burning of the fossil fuels that have been stored in the earth since it's inception, is something that has never happened to our little planet before. Once enough of the Carbon gasses are released into the atmosphere and the Glaciers start to form, there is very little that can be done in the short term and maybe even in the long. An ice age lasts for many millions of years.

When enough of the carbon gasses are in the atmosphere, the earths weather systems will start to change as they are now. The change in the worlds weather systems with an increase in atmospheric moisture will start the building of the great glaciers. When the great glaciers form then the continents will become uninhabitable. This in the past has destroyed vegetation, and that destroys animal life that depends on the vegetation. If we do not act now and learn to control our environment on an global scale, then the grim and terrible series of events that have been portrayed in this journal will certainly take place. It will not take place in fifty thousand years or longer, it will take place in less than fifty. The record cold temperatures that we are experiencing this last few winters, along with the more frequent record breaking ice and snow storms that are responsible for the increased flooding taking place in the mid west and all over the world, are ominous warnings of what is coming

If the coming ice age and the destruction it will cause is allowed to take place then it will take maybe a hundred million years for the heaver carbon gasses to settle back down to the lower altitudes over the oceans of the world. Here the simple algae that grow in the sea can metabolize them out and this may never happen. It may never happen because of the great revolving storms that the earth now must endure because of the change in the position of the continents. When the continents were a single flat mega continent there may have been less and maybe even none of these great revolving storms we call Hurricanes, Cyclones, and Monsoons. With the continental division that we have today as the Americas in the west, and Europe, Asia and Africa, in the east, these great revolving storms may keep the atmosphere with the carbon gasses in them that we put there mixed until our sun becomes a red giant star about five billion years from now. When this happens, the whole earth itself will be engulfed by the expanding sun and be destroyed.

If the carbon gasses settle down into the lower atmosphere then the simple algae's that grow in the sea can metabolize these gasses back out of the atmosphere but it will still take eons for this to happen if it does. The earths atmosphere and surface will slowly cool down, and then the Glaciers will melt and recede. This could after many millions of years restore the earth back to what we have to day.

The first and most obvious thing to due, is to plant vegetation were ever it can be planted any place on earth. Plant vegetation and plants where the sources of carbon gasses are. Keep planters with plants in them right in your house, on the porch, and out in the yard. This would help if the entire human race were to do this. If every human being on earth kept maybe a half dozen indoor plants in their house it would make a difference. That would be about thirty five to forty billion plants. These plants do not get eaten, burned for fuel, or used for building material. All they do is metabolize the carbon gasses out of the atmosphere. .Your house or apartment would smell so much fresher, cleaner, and you would be much healthier. This is because the plants are taking the carbon dioxide out of the air that you are putting there by your breathing. When they do this, they are at the same time excreting oxygen back in to the room for you to breath again. This is truly natures natural air conditioner. Put potted plants in all the office buildings of all companies in every country on earth. Put them, in all factories, in schools, in shopping malls, and yes in all government building for every government on earth. Put potted plants by the trillions in every nook an corner of our society where people and animals are. Plant trees any place that they can be planted including along highways, freeway, and rural roads and plant them by the billions all over the world.

Did you ever wonder why people like to go to the park, or get out into the country. Go camping and get out into the woods. Why it seems so nice to spread a blanket out on the grass in the park and have a picnic. When you are in a park or the forest so to speak the trees, grass, and bushes there are constantly removing the carbon gasses from the atmosphere and releasing oxygen. The air in such places is rich with a high degree of oxygen for you to breathe and this promotes a feeling of well being.

Mankind in our stupidity has destroyed most of the great forest of the world. The British isles before the great days of sail had some of the finest oak forest found anywhere on earth. Oak made beautiful sailing ships, wood flooring for homes and furniture. It was a very strong heavy wood and many ships of solid oak were built including the entire Royal British Navy. The demand for wooden sailing ships during the great days of sail decimated not only the British isles, but American and European forest as well. All over the world mankind has decimated one great forest after another. Sitka is a village on the west coast of Canada in north America that had some of the most desirable straight grained Spruce that could be found any where on earth. It commanded a premium price and was termed Sitka Spruce. Today it has all been logged of and there is no Sitka Spruce available for anything. Philippine mahogany from the Philippines is almost gone, the great Teak forest from Indochina is nearing an end, needless to say of the great tracks of Douglas Fir, Cedar, Hemlock, Poplar, Pine, Alder, Birch, and the great Red Wood forest of California are all but gone. These huge Red Wood trees many of which started to grow before the birth of Christ over 2,000 years ago in just a few short years (about 60) are almost non existent. All these great stands of timber all over the world were twenty four hours a day, seven days a week, fifty two weeks a year, year in and year out were removing carbon dioxide from the atmosphere and excreting oxygen back into it at the same time. Now they are not there any more.

Because of this do not listen to municipalities about conserving water. That is one of the biggest problem. Plants and trees need water, lots of water and we have to give it to them if we are to survive. This means plant everything and anything that will grow especially trees and shrubs. We have to try to rebuild the great forest of North America, Europe, Asia, and the African continent or else plant replacement foliage. We have to stop the destruction of the rain forest in the Amazon basin that is in south America that we have so stupidly destroyed and are still destroying. If we do not stop the destruction of the forest then we have to at least require that they plant replacement foliage that will be equivalent to what was destroyed . There is no time to waste. We have to do these things now and not only now but we have to do this FAST, FAST, FAST and FASTER. Vegetation of any kind especially TREES AND PLANTS are natures only way of effectively removing the Carbon gases from the atmosphere maintaining the necessary balance between the two main life forms. We have to learn to help nature and the natural process maintain the necessary balance and prevent another great killer ice age from developing.

There will have to be global regulation put forth by the United Nations that prohibit paper companies from any nation from using logs in the production of paper all over the world. Instead they must send crews with portable chippers and trucks into the forest. Have them remove all the under brush around and under the trees and run it through the chippers and then truck it back to the paper companies for use as paper products. It would be a good source of material for them to use. Do this instead of cutting trees and using logs for paper. By doing this you can stop making so many plastic products and make more paper products that can be recycled. By removing the underbrush in the forests world wide, lighting when it strikes will not have fuel that starts random forest fires like it does today. Forrest fires send billions of tons of carbon gas into the atmosphere every year on a global scale. Not only this but it kills the very vegetation that removes the carbon gasses from the atmosphere that promote global warming in the first place.

Again have the United Nations outlaw the clear cutting of timber any where on earth. Require timber companies again all over the world to thin the trees in a forest when they log, leaving the most productive timber to spread huge limbs and leaves. The more leaves on a tree, the better the conversion of carbon dioxide to oxygen. Require immediate refor- estation when any logging operation is performed. In the process of reforestation, plant the evergreens like fir, cedar, spruce, and hemlock. Replant millions of the giant red woods back in central and northern California. The evergreens classified as a soft wood keep their leaves all year round taking carbon out of the atmosphere year round placing oxygen back in where as the so called hard woods loose there leaves every season in winter and enter a dormant state. Have research groups do studies on all vegetation that will determine which trees and vegetation are best at removing the carbon gasses and in the process return oxygen back into the atmosphere. Then plant this type of foliage in designated areas and it is pro- tected from destruction by law.

Another thing would be to stop the building of power plants that use fossil fuels unless they can ef- fectively remove most of the carbon gasses from their emissions. Require all existing power plants that burn fossil fuels to immediately install facilities that remove most of the carbon gasses from their emis- sions. They are working on the problem right now that is evident but they need to realize the extreme urgency and what will happen if they do not slow and then stop the carbon emission altogether. This country can not do it alone so the rest of the world has to be educated to the dangers that lie a head if we do not.

Why not experiment with building covered nurseries in the vicinity of the power plants that the carbon gasses can be used for the production of plants and raise them as food. These facilities would feed the carbon gasses from the power plants right to the plants and the plants should increase their growth rate dramatically. Use a system of piping to funnel the carbon gasses right to the vegetation. Carbon gasses are to plants the same as oxygen is to people and animals. Carbon gases are an essential ingredient in the growth of plant life. The growth rate in such a facility could double, triple, and even quadruple. You could raise tomatoes, potatoes, any vegetation, fruits and vegetables that would double and even maybe triple in size and weight simply by directing the carbon gasses directly to the vegetation instead of up into the atmosphere through a great big tall smoke stack.

Why not plant large areas of tree saplings around the power plants that would be another way of disposing of and at the same time utilizing the carbon gasses. Plant hundreds of thousands of acres of young tree saplings around the plants. Again install a piping system that feeds the carbon gasses from the power plant right to the base of the trees. The trees metabolize the heavier carbon gasses, excrete the lighter oxygen and this would again promote increased growth of the trees . If a coal fired power plant was built with a thirty to a fifty year projected life and surrounded by several hundred thousand acres of new forest, by the time the life of the plant was reached, the timber would be matured. When the useful life of the old plant expired, build another new plant at a new location with a new forest of saplings that would grow with the use of the plant. Why not build these plants on federal land in an effort to restore the great forests that mankind has so stupidly destroyed

We are doing it all the wrong way by building super big high smoke stacks at power plants that carry these gasses to high altitudes where they do the most damage to the atmosphere. These gasses have to be taken right to the vegetation that will metabolize it and they in turn will produce more oxygen to balance the atmosphere.

Change the building codes so that all new homes are to have the exterior walls made of cement blocks or reinforced concrete with four of the interior wall at ninety degree to the exterior walls of the same construction. All floors in new homes are to be poured cement slabs. All are to be reinforced with rebar. In earth quake and areas prone to tornadoes and hurricanes they are to have at least one inch diameter rebar driven into the ground four feet at specific location of the exterior walls to anchor the house down to the earth. Stop the building of roof trusses made of wood and make them of steel that are either welded or bolted to the rebar used in the construction of the exterior walls of all new homes.

This right there would do several important things. It would slow the destruction of forest products that metabolize the carbon gasses from the atmosphere. It would save homes all across the country and even around the world that experience tornado and hurricane active from the massive destruction that they now must endure. Homes constructed this way would be almost indestructible when subjected to the power of a tornado, a hurricane, or a fire storm.

Conventional stucco homes look nice, are easy to build, but are instead highly inflammable death traps. A stucco house has as much wood in the construction as a home with conventional wood siding. After the framing of the house is complete and sheathed with wood, it is finished with a layer of highly inflammable Styrofoam, chicken wire and then a thin layer of stucco. The stucco itself is laced with inflammable fiber glass strands as reinforcement for the stucco to keep it from cracking. They are fire traps in high winds as in the fire storms like they have in California and now in many parts of the world. When the exterior stucco does cracks from the heat of the fire, the fire gets into the highly inflammable Styrofoam that is under the stucco and over the bone dry wood sheeting of the house. It then races through the Styrofoam under the stucco and over the dry wood sheathing and the house is gone up in flames in minutes, some times even in seconds . It is stupid to build houses like that even criminal. Not only houses but many commercial construction projects are built the same Styrofoam fire trap way.

All new homes are to have installed solar powered electric systems in states where it is feasible such as most of the southern states in the US. This would reduce the demand for electric power from the power grid and reduce the demand for more new power plants that burn coal. Germany and the German nation is on the right track and is a leader in solar powered electric systems.

Make it a major priority any where on earth to immediately extinguish all forest fires instead of letting them burn out on there own. Forest fries by them selves place untold amounts of carbon gasses directly into the atmosphere each year and need to be tightly controlled. Controlling all forest fires on a global scale would do as much good as controlling the carbon emissions from the worlds power plants and would be easier to do.

The final word or you say the bottom line is, that we are not the primary life form on the earth. We think we are but in reality we are not. The reason we are the secondary life form is because we can not produce any food. The entire human race and all animal life on the face of the earth can not produce any food of any kind. period!! Nothing!!! All we and all animal life can do is consume. Because of this we are instead the secondary life form. Plant life is the primary life form because only the plants can grow and form as food for us to consume. With out the plants and what they do, we would all perish. Our whole existence is completely depended on Vegetation and we have decimated it. If the entire human race and all animal life on earth did not have the plants to consume either directly or indirectly as food we would all die in a very short period of time.

If the entire human race and all animal life on the earth did not have the plants that remove the carbon gasses from the atmosphere and replace it with oxygen we would all die again. Over a prolonged period of time if humanity and all animal life could survive with out the plants as food, we would still all eventually die because like the closed air tight room discussed earlier, we would deplete the oxygen from the atmosphere and replace it with carbon dioxide and it would kill us all. The atmosphere of the earth is a closed room that we exist in and there has to be a balance of the less than one half of one percent of the atmospheric gasses. When combustion (any type of fire) and animal life become overly dominant destroying to much plant life then the carbon gasses build up and there will be a period of global warming that leads to an ice age as has been describe in this journal.

A great killer ice age would kill most of the life on the continents of the earth both plant and animal. When it gets below freezing plant life will not grow. Some plants leave seeds and nuts to reproduce and then they die at the first heavy frost. Other plants enter a dormant state, and loose there leaves in a seasonal cycle. They grow new leaves every spring and the cycle repeats its self but they can not grow under or on the ice of a great Glacier.

Two life forces exist on the continental plates that we live on here on our earth. These two life forces control the less than one half of one percent of the atmospheric gasses just by their existence alone. This again is plant life, animal life, and the combustion of plant life that includes you and me as part of the same system. If we do not recognize what the problem is and do something about it, then all plant life and animal life are in for a horrific Ice age. This ice age will not happen in fifty thousand years or more, it will happen tomorrow in geological time. (ten to fifty years). This Ice age will be the granddaddy of them all with glaciers maybe two miles high covering most of the continents on the earth and it will happen with frightening speed. Those huge high energy moisture laden STORMS will roll in one after another nonstop and there will not be a thing that we can do about it except try to survive as the glaciers form. . The reason should be obvious. The reason is because we are not only destroying vegetation the likes of which the earth has never seen before, (logging and cutting down forest, slash burning, clearing land for new homes, clearing land to build free ways and enlarge cities, paving land with asphalt etc.) but we are digging up mountains of fossil fuels that have lain dormant for billions of years and burning them as well that releases billions of tons of carbon gasses in to the atmosphere. If we are to survive, this simply has to change.

ABOUT THE AUTHOR

Vern G. Rickey is an Astronomer(Amateur) who resides in the state of Arizona and is engaged in the Astro-imaging of deep sky celestial phenomena. He is a retired senior citizen and has two private astronomical observing stations on nine acres of land in the Arizona high desert and he has been engaged in this activity for the last eleven years. He has been studying "Global Warming" extensively for the last several years and the effects it is having on climate change. This includes the melting of glaciers around the world, the Melting of the "Greenland ice pack" and the melting of large areas of Antarctica. He has also been studying the increase in the exceptionally serve weather systems that have arisen including the increased frequency of ice storms, record breaking snow storms, record breaking cold temperatures, and an increase in record breaking floods especially in the Mississippi River drainage system that are contrary to what should happen because of the effects of "Global Warming". There is also the fact that the Artic polar sea ice cap has not melted as fast as it has been predicted to.